METRIC CONVERSION HANDBOOK

by

MARVIN H. GREEN

CHEMICAL PUBLISHING CO.
New York, N. Y.
1978

Printed in the United States of America
ISBN 0-820-6035-1

Foreword

The United States of America is scheduled to undergo a major conversion from its customary system of measurement to the metric system. Of course metric units have been in use in our country for many years in many fields of endeavor. However, in a very few years metric units are to replace completely all of the customary units.

There will be problems. It will be quite expensive for all of us to make this complete conversion to the metric system. It will take a major effort to learn to understand the metric system and to use it properly. Of great help to us will be the present International System of Units because it greatly reduces the number of units we have to learn in the metric system, and because it links the many categories of measurement together into one sensible, coherent system. This book stresses the importance of the International System of Units.

The purpose of this book is to make the conversion easier for all of us.

Table of Contents

An Acknowledgement

In my opinion the National Bureau of Standards is the "last word" when it comes to measurement. This book is based primarily upon publications of the Bureau.

The Bureau joins similar governmental organizations throughout the world in making final decisions in the field of metrology, the science of measurement. A truly major accomplishment was the creation of the International System of Units (SI) with its precise definitions of critical units. The Bureau has also played a vital part in the development of many other definitions of units and concepts that have been adopted internationally. Examples include the acceptance of the standard acceleration of gravity, the definitions of the standard atmosphere, thermochemical calorie, inch, pound, etc.

A special thanks is extended to Louis E. Barbrow, Coordinator of Metric Activities for the National Bureau of Standards. He supplied me with several very valuable references including some of his own articles. He taught me about the survey units of length and area. It was his suggestion to identify the ambiguous mass-related units of force by labeling them as units of force. He advised me to emphasize the use of SI units for many categories of measurement. He further advised me in a private conversation in 1976 that it was his opinion that three non-SI metric units would gain strong acceptance: the liter, hectare, and the metric ton. The liter will likely have the symbol L in the United States and other English-speaking countries.

In 1977, the liter and metric ton were listed as units acceptable for use with SI, and the liter was given the symbol L for use in the United States. Furthermore, the hectare was recommended for use with SI in the United States.

Mr. Barbrow answered my many requests for help rapidly and completely. He even reviewed the whole book.

For the section on units of pressure, Mr. Barbrow provided me with the densities of water and mercury. The following is taken from his letter to me dated April 19, 1976:

"For the maximum density of water, my colleagues at NBS tell me that the PTB (Germany) 1971 value is the best available: 999.9720 kilograms per cubic meter at 3.98°C. Similarly, the NPL (England) 1961 determination of the density of mercury at 0°C is recognized as the best available: 13 595.080 kilograms per cubic meter."

If there are any errors in computations, they are mine. No one else checked my computations. However, I have made every effort and will continue my efforts to proofread and check the entire book to the best of my ability.

The choice of units included in the 17 categories of measurement is mine, although I did draw heavily on Mr. Barbrow's 1975 revision of an article on units of length, mass, area, and volume, published by the Bureau. Most of the other categories consist of units compounded from the basic units of length and mass and the familiar time units.

If this book helps people to make the conversion to the metric system, particularly to SI, it will have served its purpose. And eventually it will serve no other purpose than as a historical record of how complicated the field of measurement was before the metric era.

Marvin H. Green

Introduction and Guide

The primary purpose of this book is to provide accurate, detailed easy-access factors for converting to and from the United States customary and metric systems of measurement. The International System of Units is discussed in detail. It is basically a metric system that links the many categories of measurement together in one sensible, coherent system of units. It is properly abbreviated SI after the French Système International d'Unités, an abbreviation that will become familiar to everyone. Suggestions are made for easing the conversion from the customary system to SI, and most, but not quite all, of the units recommended for practical and scientific use are part of SI. In fact, a great many non-SI metric units should be replaced by SI units.

Conversion factors are provided for the following 17 categories of measurement:

1. Angular Measure	10. Light
2. Area	11. Mass
3. Atomic Energy Units	12. Power
4. Density and Concentration	13. Pressure
5. Electricity and Magnetism	14. Temperature
6. Energy	15. Time
7. Flow	16. Velocity
8. Force	17. Volume
9. Length	

Pocket calculators are now relatively inexpensive and available in many stores. Most handle 8- or 10-digit numbers, and therefore the majority of the conversion factors in this book are presented with 10 significant figures. All calculations were made to at least 16-digit accuracy. Of course, in most cases and for most people, such accuracy will not be needed. However, it is easy to round off a 10-digit number to the desired number of digits. It requires a great deal of work to expand the number of digits from a few to many.

For those who require and demand great accuracy, this book provides the necessary factors. With so many excellent calculators available, there may develop a desire to expend the little extra effort required to handle 10-digit numbers.

1

The tables of conversion factors comprise the major portion of the book. Major features of the tables and a guide to their use are listed below:

1. The tables are separated from the text so that the reader may get to the conversion factors as easily and rapidly as possible.

2. The 17 categories are presented in alphabetical order.

3. Within each category (except temperature) units are listed in ascending order, and under each unit the conversion factors are also presented in ascending order.

4. To avoid any ambiguity, each table is headed by a statement like the following:

One KILOGRAM-FORCE PER SQUARE CENTIMETER is equal to:

5. All computations were made to at least 16-digit accuracy, and many were calculated beyond this point to make sure that factors were or were not exact.

6. All exact relationships are presented with only the number of digits required; unnecessary zeroes are not added. For example, one calorie is equal to exactly 4.184 joules and is presented as such.

7. All exact relationships are followed by an asterisk (*) and are presented without using the powers of 10.

8. Almost all inexact relationships are presented with 10 significant figures. Exceptions are factors for temperature, and for atomic energy units and pressure, which are based on measured values with less than 10-digit accuracy.

9. If more than 10 digits are required to show that a conversion factor is exact, the factor is presented with the required number of digits. For example, the bushel is shown to equal exactly 35.239 070 166 88 liters.

10. Numbers on either side of the decimal point are printed with a space separating each group of 3 digits. Commas are not used to separate numbers to the left of the decimal point. Commas were designated to be eliminated as far back as 1948 by international agreement. In some countries the comma is used instead of the decimal point.

11. All inexact factors between 0.1 and 0.000 001 are given in decimal form. Smaller factors use exponents of 10. Those inexact factors with more than 9 digits to the left of the decimal point are also presented using the powers of 10.

12. Abbreviations are used only to save space.

13. Each section except temperature concludes with a list of additional units that are generally not used as often as those in the tables. Electricity and magnetism, and light list additional categories as well as units.

14. The section on temperature differs in style from the other sections. It consists of equations, tables, and examples designed to provide the easiest means of converting from one scale to another.

The chapter titled "The International System of Units (SI)" is necessarily very technical. It presents the precise definitions of SI base, derived, and supplementary units written by international experts. This chapter discusses SI in general and lists many non-SI units that are accepted for use with SI and others that are not accepted. A table is included that lists the SI units and their symbols, and expressions in terms of other SI units. A second table presents the SI prefixes.

We must recognize SI as a truly great accomplishment by experts from many countries, one that has far-reaching importance for both the general public and the scientific community.

The chapter titled "Categories and Units of Measurement" is also technical. Each category is presented with the list of units included in the tables. Generally units are divided into metric and customary groups. Dimensions are presented and SI units listed and discussed.

Basic or definitive conversion factors are those like 1 mile is equal to 5 280 feet, 1 square yard equals 9 square feet, etc. This chapter provides such factors for the units in the following categories:

1. Angular Measure
2. Area
3. Length
4. Light
5. Mass
6. Time
7. Volume

In order to emphasize the importance of SI, this chapter presents only the relationships to the SI unit for all units, metric and customary, for the following categories:

1. Atomic Energy Units
2. Electricity and Magnetism
3. Energy
4. Force
5. Power
6. Pressure
7. Velocity

In two categories each unit is related to a unit that is not part of SI. For density and concentration the basic unit used is the gram per liter; the SI unit is the kilogram per cubic meter, which equals the gram per liter but is not used as frequently. For flow the basic unit used is the liter per second; the SI unit is the cubic meter per second and it is too large for practical purposes. The liter is not an SI unit, but it is used very widely and is acceptable for use with SI.

In "Categories and Units of Measurement" the conversion factors are presented following the rules used in the tables: numbers in ascending order in the metric and customary groups, mostly 10-digit numbers unless fewer or more digits are required to make the relationship exact, each exact relationship followed by an asterisk, etc.

The chapter titled "Suggestions for Easing the Conversion to SI" is the least technical chapter. Its primary purpose is to serve as a practical guide to the use of few units to replace many. Most of the units recommended for use are a part of SI, but there are exceptions.

Simplified conversion factors are presented with few significant figures, and asterisks are not used to indicate exact relationships because some of these relationships require many figures to become exact. Multiples of the primary SI units are recommended in many cases because they are more compatible numerically with the other unit in the relationship. For example, the kilometer is recommended as the replacement for the mile because it is numerically closer to the mile than is the meter, which is the primary SI unit.

This chapter concludes with a table of recommended units for each category except atomic energy units. The chapter demonstrates very clearly that in many categories a few SI units may and should be used to replace the many non-SI metric and customary units listed in the tables. For the categories listed below, the number of recommended units is presented together with the number of other units in the tables that they should replace:

1. Area: 4 replace 12
2. Density and Concentration: 1 replaces 17
3. Energy: 4 replace 13
4. Flow: 2 replace 14
5. Force: 2 replace 7
6. Length: 4 replace 12
7. Mass: 4 replace 11
8. Power: 4 replace 13
9. Pressure: 2 replace 16
10. Velocity: 2 replace 9
11. Volume: 2 replace 16

The chapter titled "Lists of Additional Units" provides details on the selection of units for the lists and the methods of presentation.

The International System of Units (SI)

The International System of Units, abbreviated SI after the French Système International d'Unités, is the name given to a single, practical, worldwide system of units for international relations, teaching, and scientific work. It was adopted by the General Conference on Weights and Measures, which governs the International Committee for Weights and Measures, which in turn supervises the International Bureau of Weights and Measures. The National Bureau of Standards represents the United States in the activities and meetings of the General Conference.

A great advantage of SI is that there is a rationalized and coherent system consisting of base units, derived units, and supplementary units. The meter, kilogram, and second are three of the base units and are called the MKS group. A previously popular group, the centimeter, gram, and second, called the CGS group, is preferably not to be used with SI except in special circumstances.

BASE UNITS OF SI

There are seven well-defined base units in SI that by convention are regarded as dimensionally independent. The following definitions are copied from SI (ref 5).

1. The METER is the length equal to 1 650 763.73 wavelengths in vacuum of the radiation corresponding to the transition between the levels $2p_{10}$ and $5d_5$ of the krypton-86 atom.

2. The KILOGRAM is the unit of mass equal to the mass of the international prototype of the kilogram.

3. The SECOND is the duration of 9 192 631 770 periods of the radiation corresponding to the transition between the two hyperfine levels of the ground state of the cesium-133 atom.

4. The AMPERE is that constant current which, if maintained in two straight parallel conductors of infinite length, of negligible circular cross section, and placed 1 meter apart in vacuum, would produce between these conductors a force equal to 2×10^{-7} newton per meter of length.

5. The KELVIN, unit of thermodynamic temperature, is the fraction 1/273.16 of the thermodynamic temperature of the triple point of water.

5

6. The MOLE is the amount of substance of a system which contains as many elementary entities as there are atoms in 0.012 kilogram of carbon-12.

7. The CANDELA is the luminous intensity, in the perpendicular direction, of a surface of 1/600 000 square meter of a blackbody at the temperature of freezing platinum under a pressure of 101 325 newtons per square meter.

The only liberties taken with the definitions as presented in SI (ref 5) are that units have been capitalized, the reference uses the spelling metre (English) instead of meter (American), either of which is acceptable, and the reference presents the definitions in italics.

It is important to note that these base units are regarded as dimensionally independent *by convention*. This certainly simplifies the derivation of dimensions of units related to the base units. Obviously the ampere and candela could be considered dependent upon other units of measurement as noted in their definitions.

DERIVED UNITS OF SI

There are of course a great number of units that may be derived from the seven base units of SI. At this point the derived units that are important in this book are dealt with in brief; more detailed treatment is provided in the chapter "Categories and Units of Measurement."

One great advantage of SI is that the various categories of measurement and their SI units are linked together in a coherent, easily understood system.

The meter is the base unit for length, the kilogram is the base unit for mass, and the second is the base unit for time. These three base units and their fundamental categories of measurement serve as the bases for nine other categories of measurement in this book.

1. Area has the dimensions: length squared (l^2). Thus the SI unit is the square meter (m^2).

2. Volume has the dimensions: length cubed (l^3). Thus the SI unit is the cubic meter (m^3).

3. Velocity has the dimensions: length per time (l/t). The SI unit is the meter per second (m/s).

4. Flow has the dimensions: volume per time (l^3/t). In this case the SI unit is the cubic meter per second (m^3/s).

5. Density and concentration are really two categories of measurement treated in combination in this book, and they have the same dimensions:

mass per volume (m/l^3). The SI unit is the kilogram per cubic meter (kg/m^3).

6. Force has the dimensions: mass times acceleration (ma) and acceleration has the dimensions: length per time squared (l/t^2) thus force equals (ml/t^2). The SI unit is the kilogram-meter per second squared $(kg \cdot m/s^2)$ which is called the newton (N).

7. Pressure has the dimensions: force per area (m/lt^2). The SI unit is the newton per square meter (N/m^2) which has recently been named the pascal (P).

8. Energy has the dimensions: force times length (ml^2/t^2). The SI unit is the newton-meter, called the joule (J).

9. Power has the dimensions: energy per time (ml^2/t^3). In this case the SI unit is the joule per second, which is called the watt (W).

The kelvin is the base unit for temperature although the Celsius scale is accepted as part of SI and is the commonly used metric scale.

Light is a form of energy that has its own system of units. There are four major categories of measurement of light included in this book. The candela (cd) is the base unit of luminous intensity. The other categories have derived SI units with the following definitions taken from ASTM (ref 6).

1. The LUMEN is the luminous flux emitted in a solid angle of 1 steradian by a point source having a uniform intensity of 1 candela.

2. The LUX is the illuminance produced by a luminous flux of 1 lumen uniformly distributed over a surface of 1 square meter.

The lumen (lm) is the SI unit of the category called luminous flux and the lux (lx) is the SI unit of the category called illuminance or illumination. The fourth category is luminance and the SI unit is the candela per square meter (cd/m^2).

There are 11 major categories of measurement of electricity and magnetism in this book. The ampere is the base unit of electric current and also of magnetomotive force. The other categories have derived SI units. The following definitions are from ASTM (ref 6).

1. The COULOMB (C) is the quantity of electric charge transported in 1 second by a current of 1 ampere.

2. The VOLT (V) is the difference of electric potential between two points of a conductor carrying a constant current of 1 ampere, when the power dissipated between these points is equal to 1 watt.

3. The OHM (Ω) is the electric resistance between two points of a conductor when a constant difference of potential of 1 volt, applied between these two points, produces in this conductor a current of 1 ampere, this conductor not being the source of any electromotive force.

4. The FARAD (F) is the capacitance of a capacitor between the plates of which there appears a difference of potential of 1 volt when it is charged by a quantity of electricity equal to 1 coulomb.

5. The HENRY (H) is the inductance of a closed circuit in which an electromotive force of 1 volt is produced when the electric current in the circuit varies uniformly at a rate of 1 ampere per second.

6. The SIEMENS (S) is the electric conductance of a conductor in which a current of 1 ampere is produced by an electric potential difference of 1 volt.

7. The WEBER (Wb) is the magnetic flux which, linking a circuit of one turn, produces in it an electromotive force of 1 volt as it is reduced to zero at a uniform rate in 1 second.

8. The TESLA (T) is the magnetic flux density given by a magnetic flux of 1 weber per square meter.

The SI unit of magnetic field strength is the ampere per meter. The eleventh category is magnetomotive force or magnetic potential difference and the SI unit is the ampere; note that the ampere is defined as a current which produces a force.

The SI expressions of the derived units of electricity and magnetism are presented in the table included in this section and are dealt with in detail in the chapter "Categories and Units of Measurement."

SUPPLEMENTARY UNITS OF SI

At this time there are only two supplementary units of SI and they may be regarded either as base units or as derived units. The following definitions are taken from SI (ref 5).

1. The RADIAN (rad) is the plane angle between two radii of a circle which cut off on the circumference an arc equal in length to the radius.

2. The STERADIAN (sr) is the solid angle which, having its vertex in the center of a sphere, cuts off an area of the surface of the sphere equal to that of a square with sides of length equal to the radius of the sphere.

Angular measure is included in this book, and the SI unit of angular measure is the radian. Although the second, minute, and degree of angular measure are not part of SI, they are important units and are widely used. They are therefore acceptable units for use with SI units.

UNITS OUTSIDE SI

An excellent and basic reference for the International System is SI (ref 5). This reference categorizes several groups of units which are outside

SI. These groups, and the units included in this book that fall into each group, are the following:

1. Units that are not part of SI but which are important and widely used: minute, hour, and day of time; degree, minute, and second of angular measure; liter, symbol L; metric ton (or tonne); and hectare.

2. Units with values expressed in SI units which must be obtained by experiment: electronvolt, unified atomic mass unit, astronomical unit, and parsec.

3. Units accepted temporarily: nautical mile, knot, angstrom, are, barn, bar, and standard atmosphere.

4. CGS units with special names that it is in general preferable not to use with SI: erg, dyne, gauss, oersted, maxwell, stilb, and phot.

5. Other units generally deprecated: metric carat which is also known simply as the carat, torr, kilogram-force, International Steam Table calorie, stere, the gamma which is related to the kilogram, the gamma which is related to the tesla, fermi, and micron.

Of particular interest at this point are the units with values which must be obtained by experiment. The astronomical unit and the parsec are included in this book as additional units of length. The electronvolt and the unified atomic mass unit are included in the category, atomic energy units, although the latter is included in terms of its equivalent energy. The following definitions are taken from SI (ref 5).

1. The ELECTRONVOLT (eV) is the kinetic energy acquired by an electron in passing through a potential difference of 1 volt in vacuum.

2. The UNIFIED ATOMIC MASS UNIT (u) is equal to the fraction $1/12$ of the mass of an atom of the nuclide ^{12} C.

The following statement appears only as a footnote on page 16 in SI (ref 5), but it is quoted here because of its clarity of expression:

"The aim of the International System of Units and of the recommendations contained in this document is to secure a greater degree of uniformity, hence a better mutual understanding of the general use of units. Nevertheless in certain specialized fields of scientific research, in particular in theoretical physics, there may sometimes be very good reasons for using other systems or other units."

TABLE OF SI UNITS

The categories of measurement listed in the table of SI units presented below include some groups. In some cases the grouping means that the categories have the same dimensions and units, as in the case of density and concentration. Work and quantity of heat are grouped with energy because they are both forms of energy. On the other hand, illuminance and illumination are grouped because they are different names for the same category, although illuminance is preferred.

The symbols are from SI (ref 5).

Expressions in terms of SI base units are presented using exponents although some references prefer using negative exponents in lieu of the solidus (/). A dot should be used to indicate the product of two or more units but may be dispensed with when there is no risk of confusion with another unit sysmbol: N·m or N m means newton times meter, but mN means millinewton.

Category of Measurement	Name of SI Unit	Symbol of SI Unit	Expression in Terms of Other SI Units	Expression in Terms of SI Base Units
length	meter	m		m
mass	kilogram	kg		kg
time	second	s		s
area	square meter	m^2		m^2
volume	cubic meter	m^3		m^3
velocity, speed	meter per second	m/s		m/s
flow	cubic meter per second	m^3/s		m^3/s
density, concentration	kilogram per cubic meter	kg/m^3		kg/m^3
force	newton	N		$kg \cdot m/s^2$
pressure, stress	pascal	Pa	N/m^2	$kg/(m \cdot s^2)$
energy, work, quantity of heat	joule	J	$N \cdot m$	$kg \cdot m^2/s^2$
power, radiant flux	watt	W	J/s	$kg \cdot m^2/s^3$
angular measure, plane angle	radian	rad		rad
thermodynamic temperature	kelvin	K		K

Category of Measurement	Name of SI Unit	Symbol of SI Unit	Expression in Terms of Other SI Units	Expression in Terms of SI Base Units
luminous intensity	candela	cd		cd
luminous flux	lumen	lm		cd·sr
luminance	candela per square meter	cd/m^2		cd/m^2
illuminance, illumination	lux	lx	lm/m^2	$cd·sr/m^2$

Category of Measurement	Name of SI Unit	Symbol of SI Unit	Expression in Terms of Other SI Units	Expression in Terms of SI Base Units
electric current	ampere	A		A
electric charge, quantity of electricity	coulomb	C		A·s
electric potential, potential difference, electromotive force	volt	V	W/A	$kg·m^2/(s^3·A)$
electric resistance	ohm	Ω	V/A	$kg·m^2/(s^3·A^2)$
electric capacitance	farad	F	C/V	$s^4·A^2/(kg·m^2)$
electric inductance	henry	H	V·s/A	$kg·m^2/(s^2·A^2)$
electric conductance	siemens	S	A/V	$s^3·A^2/(kg·m^2)$
magnetic flux	weber	Wb	V·s	$kg·m^2/(s^2·A)$
magnetic flux density, magnetic induction	tesla	T	Wb/m^2	$kg/(s^2·A)$
magnetic field strength	ampere per meter	A/m		A/m

Category of Measurement	Name of SI Unit	Symbol of SI Unit	Expression in Terms of Other SI Units	Expression in Terms of SI Base Units
magnetomotive force, magnetic potential difference	ampere	A		A
amount of substance	mole	mol		mol
solid angle	steradian	sr		sr

SI PREFIXES

The table below presents the prefixes to be used with units of SI. Barbrow is the source of the prefixes exa and peta, and their symbols, in his 1975 revision of the article written by Judson in 1960: Barbrow and Judson (ref 1).

The names of the numbers are not usually given in official documents. The names in the table are those used in the United States where the billion means 1 000 million. The British billion equals 1 000 000 million, and 1 000 million is the British milliard.

Factor	Prefix	Symbol		Name
10^{18}	exa	E	1 000 000 000 000 000 000	quintillion
10^{15}	peta	P	1 000 000 000 000 000	quadrillion
10^{12}	tera	T	1 000 000 000 000	trillion
10^{9}	giga	G	1 000 000 000	billion
10^{6}	mega	M	1 000 000	million
10^{3}	kilo	k	1 000	thousand
10^{2}	hecto	h	100	hundred
10^{1}	deka	da	10	ten
10^{-1}	deci	d	0.1	tenth
10^{-2}	centi	c	0.01	hundredth
10^{-3}	milli	m	0.001	thousandth
10^{-6}	micro	μ	0.000 001	millionth
10^{-9}	nano	n	0.000 000 001	billionth

Factor	Prefix	Symbol		Name
10^{-12}	pico	p	0.000 000 000 001	trillionth
10^{-15}	femto	f	0.000 000 000 000 001	quadrillionth
10^{-18}	atto	a	0.000 000 000 000 000 001	quintillionth

Categories and Units of Measurement

This chapter lists all of the categories of measurement covered in this book and presents the units included in the tables for each category. In most cases units are listed as either metric or customary. The metric units include SI units, but not all metric units are part of SI. For example, centimeter-gram-second or CGS units are preferably not used with SI. Some meter-kilogram-second or MKS units are generally deprecated, including the kilogram-force.

Customary units are those in general use in the United States and include such units as the inch, pound, gallon, acre, mile per hour, etc.

The first three categories listed are length, mass, and time. The next group of nine categories have dimensions based on length, mass, and time. These are followed by angular measure, temperature, light, electricity and magnetism, and atomic energy units. Following is the order of presentation of the categories of measurement:

1. Length
2. Mass
3. Time
4. Area
5. Volume
6. Velocity
7. Flow
8. Density and Concentration
9. Force
10. Pressure
11. Energy
12. Power
13. Angular Measure
14. Temperature
15. Light
16. Electricity and Magnetism
17. Atomic Energy Units

Two fundamental relationships between SI and the customary system are the following:

1 inch = 0.025 4 meter, exactly
1 pound = 0.453 592 37 kilogram, exactly
In this book the pound used without an adjective means the avoirdupois pound. The troy or apothecaries pound is generally used only as a unit of mass and that not to a great extent.

1. LENGTH

Length is a fundamental category, symbol l.
The following is quoted from Barbrow and Judson (ref. 1):

"Since 1959 the yard is defined as being equal exactly to 0.9144 meter; the new value is shorter than the old value by two parts in a million. At the same time it was decided that any data expressed in feet derived from geodetic surveys within the U.S. would continue to bear the relationship as defined in 1893 (one foot equals 1200/3937 meter). This foot is called the U.S. survey foot, while the foot defined in 1959 is called the international foot. Measurements expressed in survey miles, survey feet, rods, chains, links, or the squares thereof, and also acres should therefore be converted to the corresponding metric values by using pre-1959 conversion factors where more than five significant figure accuracy is involved."

Therefore, this book includes the survey mile, survey foot, rod, chain, and link as units based on the 1893 definition of the foot. In addition, the international foot and mile are included but without the adjective. From the quoted statements it appears that the yard is not related to geodetic surveys.

Following is the list of the units of length presented in the tables with the metric units related to the SI unit of length, which is the meter:

Metric

1 angstrom	=	0.000 000 000 1 . . . . . meter*
1 micrometer	=	0.000 001 . . . . meter*
1 millimeter	=	0.001 . . . . meter*
1 centimeter	=	0.01 . . . . meter*
meter		
1 kilometer	= 1 000	. . . . meters*
1 nautical mile	= 1 852	. . . . meters*

Customary

inch			
1 link	=	0.66	. . . . survey foot*
1 foot	=	12	. . . . inches*
1 survey foot	=	1.000 002 000	. . . . feet
1 yard	=	3	. . . . feet*
1 rod	=	16.5	. . . . survey feet*
1 chain	=	4	. . . . rods*
1 mile	=	5 280	. . . . feet*
1 survey mile	=	5 280	. . . . survey feet*

2. MASS

Mass is a fundamental category, symbol *m*.

Following is the list of the units presented in the tables with the metric units related to the SI unit of mass, which is the kilogram:

Metric

1 milligram	=	0.000 001	. . . kilogram*
1 gram	=	0.001	. . . . . . kilogram*
kilogram			
1 metric ton	=	1 000	. . . . kilograms*

Customary

grain			
1 apothecaries scruple	=	20	. . . . grains*
1 pennyweight	=	24	. . . . grains*
1 avoirdupois dram	=	27.343 75	. . . . grains*
1 apothecaries dram	=	3	. . . . apothecaries scruples*
1 avoirdupois ounce	=	16	. . . . avoirdupois drams*
1 apothecaries or troy ounce	=	8	. . . . apothecaries drams*
1 apothecaries or troy pound	=	12	. . . . apothecaries or troy ounces*

```
1 avoirdupois
   pound         =      16  . . . . . . . . avoirdupois ounces*
1 short ton      =  2 000  . . . . . . . . avoirdupois pounds*
1 long ton       =  2 240  . . . . . . . . avoirdupois pounds*
```

3. TIME

Time is a fundamental category, symbol t.

The units of time in the tables are the everyday units which have the familiar relationships: 1 minute equals 60 seconds, 1 hour equals 60 minutes, 1 day equals 24 hours, 1 year equals 365 days, etc. Units of time based on astronomical definitions are discussed in the list of additional units of time.

Metric and customary units of time are the same. The SI unit is the second. Units in the tables are the following:

```
second
1 minute        =   60  . . . . . . . . . . . . . . . . . . . . . seconds*
1 hour          =   60  . . . . . . . . . . . . . . . . . . . . . minutes*
1 day           =   24  . . . . . . . . . . . . . . . . . . . . . hours*
1 week          =    7  . . . . . . . . . . . . . . . . . . . . . days*
1 28-day month  =  672  . . . . . . . . . . . . . . . . . . . . . hours*
1 29-day month  =  696  . . . . . . . . . . . . . . . . . . . . . hours*
1 30-day month  =  720  . . . . . . . . . . . . . . . . . . . . . hours*
1 31-day month  =  744  . . . . . . . . . . . . . . . . . . . . . hours*
1 year          =  365  . . . . . . . . . . . . . . . . . . . . . days*
1 leap year     =  366  . . . . . . . . . . . . . . . . . . . . . days*
```

4. AREA

Area has the dimensions length squared, symbol l^2. The SI unit is the square meter. Included are the square link, square survey foot, square rod, square chain, acre, and square survey mile, which are all based on the 1893 definition of the foot. Also included are the square foot and square mile which are based on the definition of the yard in 1959.

Following is the list of the units presented in the tables with the metric units related to the SI unit:

Metric

```
1 square millimeter  =            0.000 001 . . . . square meter*
```

1 square centimeter =   0.000 1 . . . . . square meter*
square meter
1 are   =   100 . . . . square meters*
1 hectare   =   10 000 . . . . square meters*
1 square kilometer   = 1 000 000 . . . . square meters*

Customary

square inch
1 square link   =   0.435 6 . . . . . square survey foot*
1 square foot   =   144 . . . . square inches*
1 square survey foot =   1.000 004 000. square feet
1 square yard   =   9 . . . . square feet*
1 square rod   =   625 . . . . square links*
1 square chain   =   16 . . . . square rods*
1 acre   =   10 . . . . square chains*
1 square mile   = 3 097 600 . . . . square yards*
1 square survey mile =   640 . . . . acres*

5. VOLUME

Volume has the dimensions length cubed, symbol l^3. The SI unit is the cubic meter. The liter is not part of SI, but it is widely used and is included in the tables. The cubic decimeter is equal to 1 liter and is also included. The cubic centimeter is included as well as the milliliter to which it is exactly equal. Thus there are three SI units each with the adjective cubic. The kiloliter, which is equal to the cubic meter, is included to provide the third non-SI metric unit to go with the liter and milliliter.

Customary units of liquid and dry measures are linked to the cubic inch by the following relationships:

1 gallon   =   231   cubic inches, exactly
1 bushel   = 2 150.42 cubic inches, exactly

Following is the list of the units presented in the tables with the metric units related to the SI unit:

Metric

1 cubic centimeter   =   0.000 001. . . . . cubic meter*
1 milliliter   =   0.000 001. . . . . cubic meter*
1 cubic decimeter   =   0.001 . . . . cubic meter*

1 liter	=	0.001 . . . . cubic meter*
cubic meter		
1 kiloliter	=	1 . . . . cubic meter*

Customary

Volume
  cubic inch

| 1 cubic foot | = | 1 728 . . . . cubic inches* |
| 1 cubic yard | = | 27 . . . . cubic feet* |

Liquid Measure
  minim

1 fluid dram	=	60 . . . . minims*
1 fluid ounce	=	8 . . . . fluid drams*
1 gill	=	4 . . . . fluid ounces*
1 liquid pint	=	4 . . . . gills*
1 liquid quart	=	2 . . . . liquid pints*
1 gallon	=	4 . . . . liquid quarts*
1 petroleum barrel	=	42 . . . . gallons*

Dry Measure
  dry pint

1 dry quart	=	2 . . . . dry pints*
1 peck	=	8 . . . . dry quarts*
1 bushel	=	4 . . . . pecks*

6. VELOCITY

Velocity has the dimensions length per time, symbol l/t. Speed has the same dimensions. The SI unit is the meter per second.

The knot is equal to 1 nautical mile per hour, which is equal to 1.852 kilometers per hour.

Following is the list of the units presented in the tables with all units, metric and customary, related to the SI unit with which all of the customary units have exact relationships:

Metric

1 centimeter per		
second	=	0.01 . . . . .meter per second*
1 meter per minute	=	0.016 666 666 67 . . .meter per second

1 kilometer per hour	=	0.277 777 777 8 . . . .meter per second
1 knot	=	0.514 444 444 4 . . . .meter per second
meter per second		
1 kilometer per minute	=	16.666 666 67 . . . . .meters per second

Customary

1 foot per minute	=	0.005 08 . . . . .meter per second*
1 foot per second	=	0.304 8 . . . . .meter per second*
1 mile per hour	=	0.447 04 . . . . .meter per second*
1 mile per minute	=	26.822 4 . . . . .meters per second*
1 mile per second	= 1	609.344 . . . . .meters per second*

7. FLOW

Flow has the dimensions: volume per time, symbol l^3/t. The SI unit is the cubic meter per second, however, this unit is too large for most practical purposes and the liter per second is used as the basic unit of flow in this book.

Following is the list of the units presented in the tables with all units related to the liter per second with which all of the customary units have exact relationships:

Metric

1 cubic centimeter per second	=	0.001 . . . . .liter per second*
1 milliliter per second	=	0.001 . . . . .liter per second*
1 cubic decimeter per minute	=	0.016 666 666 67 . . .liter per second
1 liter per minute	=	0.016 666 666 67 . . .liter per second
1 cubic decimeter per second	=	1 . . . . .liter per second*
liter per second		
1 cubic meter per minute	=	16.666 666 67 . . . . . .liters per second

```
1 kiloliter
   per minute     =      16.666 666 67 . . . . . .liters per second
1 cubic meter
   per second     = 1 000  . . . . . . . . . . . . . .liters per second*
1 kiloliter
   per second     = 1 000  . . . . . . . . . . . . . .liters per second*
```

Customary

```
1 petroleum barrel
   per hour       =      0.044 163 137 48 . . .liter per second*
1 gallon per minute =   0.063 090 196 4 . . . .liter per second*
1 cubic foot per
   minute         =      0.471 947 443 2 . . . .liter per second*
1 gallon per second =   3.785 411 784 . . . . . .liters per second*
1 cubic yard per
   minute         =      12.742 580 966 4 . . . .liters per second*
1 cubic foot per
   second         =      28.316 846 592 . . . . . .liters per second*
```

8. DENSITY AND CONCENTRATION

Density and concentration are two categories of measurement that have the same dimensions: mass per volume, which is equal to mass per length cubed. The SI unit is the kilogram per cubic meter. For the sake of brevity, the gram per liter is used as the basic unit in this book; 1 gram per liter equals 1 kilogram per cubic meter. Both units are included in the tables.

Following is the list of the units presented in the tables with all units related to the gram per liter:

Metric

```
1 gram per cubic
   meter          =      0.001 . . . . . . . . . . . . gram per liter*
1 gram per kiloliter =   0.001 . . . . . . . . . . . . gram per liter*
gram per liter
1 kilogram per
   cubic meter    =      1 . . . . . . . . . . . . . . . gram per liter*
```

1 gram per cubic centimeter	=	1 000	grams per liter*
1 gram per milliliter	=	1 000	grams per liter*

Customary

1 grain per cubic foot	=	0.002 288 351 911 . .	gram per liter
1 grain per gallon	=	0.017 118 061 05 . . .	gram per liter
1 pound per cubic yard	=	0.593 276 421 3 . . .	gram per liter
1 ounce per cubic foot	=	1.001 153 961	grams per liter
1 grain per cubic inch	=	3.954 272 101	grams per liter
1 ounce per gallon	=	7.489 151 707	grams per liter
1 pound per cubic foot	=	16.018 463 37	grams per liter
1 pound per gallon	=	119.826 427 3	grams per liter
1 short ton per cubic yard	=	1 186.552 843	grams per liter
1 long ton per cubic yard	=	1 328.939 184	grams per liter
1 ounce per cubic inch	=	1 729.994 044	grams per liter
1 pound per cubic inch	=	27 679.904 71	grams per liter

9.   FORCE

Force has the dimensions: mass times acceleration, and acceleration has the dimensions: length per time squared. Thus:

$$\text{force} = \text{mass} \times \text{acceleration} = m \times \frac{l}{t^2} = \frac{ml}{t^2}$$

The acceleration of gravity, symbol g, is a factor in five of the units of force in this section. The standard acceleration of gravity equals 9.806 65 meters per second squared, which is equivalent to approximately 32.174 05 feet per second squared. Observed values of g differ by over 0.5 percent at various points on the earth's surface, according to Page (ref 4).

The kilogram-meter per second squared is the SI unit of force and it is called the newton. In the CGS group the basic unit is the gram-centimeter per second squared, called the dyne. The basic customary unit of force is the poundal, which is equal to one pound-foot per second squared.

Unfortunately some units of force in general use have the same names as the units of mass from which they are derived. In this book there are five such units and they appear in the units of energy, power, and pressure as well as in force. In all cases where they appear they are clearly identified as units of force to avoid any confusion with similarly named units of mass. The five units are the following:

gram-force	pound-force
kilogram-force	short ton-force
metric ton-force	

There is some confusion about the term "weight" as it relates to force and mass. In the use of the term in our everyday lives, weight has the connotation of the quantity mass. It cannot be expected that people will change their habitual use of the word, and diets will still be undertaken to lose weight, not mass. A clear explanation of the proper use of the terms is presented by Page (ref 4):

> "Since weight cannot be given a precise definition except under very restricted conditions, it is not a primary technical term. The best resolution of the weight/mass dilemma is to avoid the term 'weight' in any strict technical sense, to use 'force' and 'mass' when the distinction is important, and allow the use of 'weight' as a common synonym for mass in nontechnical material."

It is strongly recommended that, as the metric system comes into broader use, the newton should be used as the primary unit of force. Units such as the gram-force, kilogram-force, metric ton-force, pound-force, and short ton-force should not be used. Following are the relationships between the units of force and mass:

1 gram-
force   =   1. . . . . .gram mass $\times$ g
        =   980.665. . . . . .gram mass-centimeters per
                                          second squared*
        =   980.665. . . . . .dynes*

1 kilo-
gram-
force = 1 . . . . kilogram mass × g

= 9.806 65 . . . . kilogram mass-meters/second
squared*

= 9.806 65 . . . . newtons*

1 metric
ton-force = 1 000 . . . . kilograms-force = 9 806.65
newtons*

1 pound- = 1 . . . . pound mass × g ×
force

$$\frac{0.453\ 592\ 37\ \text{kilogram mass}}{\text{pound mass}}$$

= 0.453 592 37 . . . . kilogram mass ×

$$\frac{9.806\ 65\ \text{meters}}{\text{second}^2}$$

= 4.448 221 615 260 5 . . newtons*

1 short
ton-force = 2 000 . . . . pounds-force*

= 8 896.443 230 521 . . . . newtons*

Following is the list of the units presented in the tables with all units related to the SI unit with which all of the customary units have exact relationships:

Metric

1 dyne	=	0.000 01 . . . . newton*
1 gram-force	=	0.009 806 65 . . . . newton*
newton		
1 kilogram-force	=	9.806 65 . . . . newtons*
1 kilonewton	= 1 000 . . . . newtons*
1 metric ton-		
force	= 9 806.65 . . . . newtons*

Customary

1 poundal	=	0.138 254 954 376 . . . . newton*
1 pound-force	=	4.448 221 615 260 5 . . . . newtons*
1 short ton-force	= 8 896.443 230 521 . . . . newtons*

10. PRESSURE

Pressure has the dimensions: force per area. Stress has the same dimensions. Thus:

$$\text{pressure} = \frac{\text{force}}{\text{area}} = \frac{ml}{t^2} \times \frac{1}{l^2} = \frac{m}{lt^2}$$

The kilogram per meter-second squared is the SI unit of pressure and it has recently been given the name pascal (Pa). The pascal is also equal to the newton per meter squared, as follows:

$$\frac{\text{newton}}{\text{meter}^2} = \frac{\text{kilogram-meter}}{\text{second}^2} \times \frac{1}{\text{meter}^2} = \frac{\text{kilogram}}{\text{meter-second}^2} = \text{pascal}$$

Included in the units of pressure are columns of water and mercury. The maximum density of liquid water occurs at 3.98°C and is equal to 0.999 972 0 gram per cubic centimeter. A one-centimeter cube of water at 3.98°C will exert a pressure of 0.999 972 0 gram-force on its base of 1 square centimeter. A column of water 1 inch high will exert a pressure 2.54 times as great, or approximately 2.539 929 grams-force on a base of 1 square centimeter.

The density of mercury at 0°C is 13.595 080 grams per cubic centimeter. Similarly, a one-centimeter cube of mercury at 0°C will exert a pressure of 13.595 080 grams-force on the square centimeter base, and a column 1 inch high will exert a pressure of about 34.531 50 grams-force on a base of 1 square centimeter.

One atmosphere is defined as equal to 101 325 pascals.

Following is the list of the units presented in the tables with all units related to the SI unit:

Metric

1 dyne per square centimeter = pascal		0.1 . . . . pascal*
1 kilogram-force/square meter	=	9.806 65 . . . . pascals*
1 gram-force/square centimeter	=	98.066 5 . . . . pascals*
1 millibar	=	100 . . . . pascals*
1 kilopascal	=	1 000 . . . . pascals*

1 metric ton-force/square meter	=	9 806.65 . . . . pascals*
1 newton/square centimeter	=	10 000 . . . . pascals*
1 kilogram-force/square centimeter	=	98 066.5 . . . . pascals*
1 bar	=	100 000 . . . . pascals*
1 atmosphere	=	101 325 . . . . pascals*

Customary

1 pound-force per square foot	=	47.880 258 98 . . . . . pascals
1 pound-force per square inch	=	6 894.757 293 . . . . pascals
1 short ton-force/square foot	=	95 760.517 96 . . . . pascals

Other Units

1 millimeter of mercury (0°C)	=	133.322 19 . . . . pascals
1 inch of water (3.98°C)	=	249.081 9 . . . . pascals
1 foot of water (3.98°C)	=	2 988.983 . . . . pascals
1 inch of mercury (0°C)	=	3 386.384 . . . . pascals

11. ENERGY

Energy has the dimensions: force times length. Work and heat are two forms of energy with the same dimensions. Thus:

$$\text{energy} = \text{force} \times \text{length} = \frac{ml}{t^2} \times l = \frac{ml^2}{t^2}$$

The kilogram-meter squared per second squared is the SI unit of energy, and it is called the joule. The joule is also equal to the newton-meter, as follows:

$$\text{newton-meter} = \frac{\text{kilogram-meter}}{\text{second}^2} \times \text{meter} = \frac{\text{kilogram-meter}^2}{\text{second}^2} = \text{joule}$$

Included in the energy units is the British thermal unit, the Btu. It is normally defined as the amount of heat required to raise the temperature of 1 pound of water 1°F. Similarly, the calorie was originally defined as the amount of heat required to raise the temperature of 1 gram of water 1°C. As thermal measurements were made with greater precision, this definition of the calorie was not sufficiently accurate. For instance, the heat required to raise 1 gram of water from 15 to 16°C is not the same as that required to raise 1 gram from 60 to 61°C.

Included in the energy units is the calorie as it is now defined: 1 calorie is equal to 4.184 joules. This is quite close to the mean calorie, which is 0.01 times the amount of heat required to raise 1 gram of water from 0 to 100°C.

In this book the Btu is related to the calorie on the basis of the original definitions, but the calorie defined as equal to 4.184 joules is the relationship used to compare the Btu to the other energy units. The relationship between the Btu and the calorie is calculated as follows, using the fact that an interval of 1°C is equal to an interval of 1.8°F in temperature:

$$\frac{1 \text{ pound-}1°F}{Btu} \times \frac{453.592\ 37 \text{ grams}}{pound} \times \frac{1°C}{1.8°F} \times \frac{calorie}{1 \text{ gram-}1°C}$$

1 Btu = 251.995 761 1   calories

The watt and kilowatt, and the metric, electric, and customary horsepower units are units of power which are converted to energy units by multiplying each by the hour.

Following is the list of the units presented in the tables with all units related to the SI unit:

Metric

1 erg	=	0.000 000 1	joule*
1 microjoule	=	0.000 001	joule*
1 gram-force centimeter	=	0.000 098 066 5	joule*
joule			
1 calorie	=	4.184	joules*
1 kilogram-force meter	=	9.806 65	joules*
1 kilojoule	=	1 000	joules*
1 watt-hour	=	3 600	joules*

1 kilocalorie	=	4 184	. . . . .joules*
1 megajoule	=	1 000 000	. . . . .joules*
1 metric horse- power-hour	=	2 647 795.5	. . . . .joules*
1 electric horse- power-hour	=	2 685 600	. . . . .joules*
1 kilowatt-hour	=	3 600 000	. . . . .joules*

Customary

1 foot poundal	=	0.042 140 110 093 804 8	. . . . . .joule*
1 foot pound- force	=	1.355 817 948 331 400 4	. . . . . .joules*
1 Btu	=	1 054.350 264	. . . . .joules
1 horsepower- hour	=	2 684 519.537 696 172 792	. . . . .joules*

12. POWER

Power has the dimensions energy per time. Radiant flux has the same dimensions. Thus:

$$\text{power} = \frac{\text{energy}}{\text{time}} = \frac{ml^2}{t^2} \times \frac{1}{t} = \frac{ml^2}{t^3}$$

The kilogram-meter squared per second cubed is the SI unit of power, and it is called the watt (**W**). The watt is also equal to the joule per second, as follows:

$$\frac{\text{joule}}{\text{second}} = \frac{\text{kilogram-meter}^2}{\text{second}^2} \times \frac{1}{\text{second}} = \frac{\text{kilogram-meter}^2}{\text{second}^3} = \text{watt}$$

The metric horsepower is defined as equal to 75 kilogram-force meters per second. The electric horsepower is defined as equal to 746 watts. The customary horsepower is defined as equal to 550 foot pounds-force per second.

Following is the list of the units presented in the tables with all units related to the SI unit:

Metric

1 erg per second	=	0.000 000 1	watt*
1 microwatt	=	0.000 001	watt*
1 gram-force centimeter/ second	=	0.000 098 066 5	watt*
1 kilogram-force meter/ minute	=	0.163 444 166 7	watt

watt

1 calorie per second	=	4.184	watts*
1 kilocalorie per minute	=	69.733 333 33	watts
1 metric horse-power	=	735.498 75	watts*
1 electric horse-power	=	746	watts*
1 kilowatt	=	1 000	watts*
1 megawatt	=	1 000 000	watts*

Customary

1 foot pound-force per minute	=	0.022 596 965 805 523 34	watt*
1 Btu per hour	=	0.292 875 073 5	watt
1 foot pound-force per second	=	1.355 817 948 331 400 4	watts*
1 Btu per minute	=	17.572 504 41	watts
1 horsepower	=	745.699 871 582 270 22	watts*
1 Btu per second	=	1 054.350 264	watts

13. ANGULAR MEASURE

Angular measure is a category which has no dimensions related to the other categories. The SI unit is the radian. Customary and metric units are the same. The grade in the tables is called the grad in some references. Units in the tables are the following:

```
second
1 minute   = 60 ............................ seconds*
1 grade    = 54 ............................ minutes*
1 degree   = 60 ............................ minutes*
1 radian   = 1/2π.......................... circle*
           =   0.159 154 943 1.............. circle
1 quadrant = 90 ............................ degrees*
1 circle   =  4 ............................ quadrants*
```

14. TEMPERATURE

Temperature is also a category which has no dimensions related to the other categories. The kelvin is the SI unit of thermodynamic temperature. Use is also made in SI of Celsius temperature.

The section on temperature includes information about scales in current use, provides equations for converting between scales, and presents five tables designed to simplify conversions between the Fahrenheit and Celsius scales.

15. LIGHT

There are four categories of measurement of light considered to be major categories in this book. The definitions of the SI units of these categories are presented in the section "The International System of Units (SI)."

(1) LUMINUOUS INTENSITY

The candela (cd) is the SI unit of the luminous intensity, and it is an SI base unit. There are no tables of units of luminous intensity because the candela is considered to be the only unit of importance.

(2) LUMINOUS FLUX

Luminous flux has the dimensions luminous intensity times solid angle. The lumen (lm) is the SI unit and it equals the candela-steradian. There are no tables of units of luminous flux because the lumen is considered to be the only unit of importance.

(3) *LUMINANCE*

Luminance has the dimensions luminous intensity per area. The SI unit is the candela per square meter. There are 7 units included in the tables and all of them are based on the candela with 4 using metric areas and the other 3 with areas in customary units. The latter 3 are labeled metric-customary units. Units in the tables are the following:

Metric

candela per square meter			
1 millilambert	=	0.001	lambert*
1 lambert	=	$1/\pi$	candela/square centimeter*
	=	0.318 309 886 2 . .	candela/square centimeter
1 stilb	=	1	candela per square centimeter*

Metric-Customary

1 footlambert	=	$1/\pi$	candela per square foot*
	=	0.318 309 886 2 . .	candela/square foot
1 candela per square foot	=	10.763 910 42	candelas/square meter
1 candela per square inch	= 1	550.003 100	candelas/square meter

(4) *ILLUMINANCE, ILLUMINATION*

Illuminance or illumination has the dimensions luminous flux per area. The SI unit is the lux (lx) which is equal to the lumen per square meter. The lumen is equal to 1 candela-steradian, and thus the lux is equal to the following expression in SI base units: candela-steradian per square meter.

There are 4 units included in the tables and 3 of them are totally metric in their definitions. The fourth, the foot-candle, is a metric-customary unit and could possibly be given the name footcandela.

Metric

1 lux	=	1	lumen per square meter*
1 milliphot	=	0.001	phot*
1 phot	=	1	lumen per square centimeter*

Metric-Customary

1 footcandle	=	1	lumen per square foot*

16. ELECTRICITY AND MAGNETISM

There are 11 major categories of measurement of electricity and magnetism in this book. The definitions of the SI units are presented in the section "The International System of Units (SI)."

The units included in the tables are all metric units. Each of the 11 categories has an SI unit. There are two CGS systems of units in use primarily for scientific purposes: the electromagnetic system which uses the prefix "ab" for many units and also uses the abbreviations EMU, and the electrostatic system which uses the prefix "stat" and the abbreviation ESU.

In SI the units of electricity and magnetism may be expressed in four SI base units: the meter, kilogram, second, and ampere. This is referred to as the MKSA group. The two CGS systems each use only three basic units, which are of course the centimeter, gram, and second. The basic relationships which link the three systems together are the following:

1 abampere	= 10	amperes*
1 abampere	= c	statamperes*

where c is the speed of light in a vacuum and has the following value from Cohen and Taylor (ref 2):

$$c = 2.997\ 924\ 58 \times 10^{10}\ .\ .\ .\ .\ .\ \text{centimeters per second}$$

Both electromagnetic and electrostatic units are presented for electricity. Electrostatic units for magnetism have no special names and are excluded as not being very important.

The oersted is the electromagnetic unit of magnetic field strength. It is equal to $1000/4\pi$ or 79.577 ampere-turns per meter. The gilbert is the electromagnetic unit of magnetomotive force, and it equals $1/0.4\pi$ or 0.795 77 ampere-turn. But there is no ampere-turn in SI, where the ampere is the unit of magnetomotive force. Therefore, the oersted corresponds to $1000/4\pi$ amperes per meter, and the gilbert corresponds to $1/0.4\pi$ ampere. The expression "corresponds to" is a better representation of the relationships than the word "equals."

The 7 categories of units of electricity and the 4 categories of units of magnetism are listed below with the units included in the tables for each category. All units are metric. Dimensions are discussed, and the SI units are expressed in terms of SI base units through the introduction of the ampere, which is the base unit of electric current, and the watt, which is the SI base unit of power. The watt is included in the definition of the volt.

(1) ELECTRIC CURRENT

The ampere is the SI base unit of electric current. The tables include only 3 metric units, as is true of the tables of the other categories of electrical units. Units in the tables are the following:

1 statampere = $3.335\ 640\ 95 \times 10^{-10}$ . . . . ampere
ampere
1 abampere = 10 . . . . amperes*

(2) ELECTRIC CHARGE, QUANTITY OF ELECTRICITY

Electric charge, which is also known as the quantity of electricity, has

$$\text{Electric charge} = \text{electric current} \times \text{time}$$
$$= It$$

the dimensions electric current times time. The SI unit is the coulomb and it is equal to 1 ampere-second. The electromagnetic unit, the abcoulomb, equals 1 abampere-second, and the electrostatic unit, the statcoulomb, is equal to 1 statampere-second. Units in the tables are:

1 statcoulomb = 3.335 640 95 × 10^{-10} . . . . .coulomb
coulomb
1 abcoulomb = 10 . . . . .coulombs*

(3) *ELECTRIC POTENTIAL, POTENTIAL DIFFERENCE, ELECTROMOTIVE FORCE*

Electric potential, also known as potential difference and electromotive

$$\text{electric potential} = \frac{\text{power}}{\text{electric current}} = \frac{ml^2}{t^3 I}$$

force, has the dimensions power per electric current. The volt is the SI unit of electric potential. In terms of SI base units, the watt, the SI unit of power, is equal to the kilogram-meter squared per second cubed. The volt equals 1 watt per ampere and may be expressed in the four SI base units as follows:

$$\frac{\text{watt}}{\text{ampere}} = \frac{\text{kilogram-meter}^2}{\text{second}^3} \times \frac{1}{\text{ampere}} = \frac{\text{kilogram-meter}^2}{\text{second}^3\text{-ampere}} = \text{volt}$$

The erg per second is the CGS unit of power. The abvolt is equal to 1 erg per second per abampere, and the statvolt is equal to 1 erg per second per statampere. Units in the tables are:

1 abvolt = 0.000 000 01 . . . . .volt*
volt
1 statvolt = 299.792 458 . . . . .volts

(4) *ELECTRIC RESISTANCE*

Electric resistance has the dimensions electric potential per electric

$$\text{electric resistance} = \frac{\text{electric potential}}{\text{electric current}} = \frac{ml^2}{t^3 I^2}$$

current. The ohm is the SI unit and equals 1 volt per ampere. The ohm may be expressed in the four base units as follows:

$$\frac{\text{volt}}{\text{ampere}} = \frac{\text{kilogram-meter}^2}{\text{second}^3\text{-ampere}} \times \frac{1}{\text{ampere}} = \frac{\text{kilogram-meter}^2}{\text{second}^3\text{-ampere}^2} = \text{ohm}$$

The abohm equals 1 abvolt per abampere and the statohm is equal to 1 statvolt per statampere. Units in the tables are as follows:

1 abohm   = 0.000 000 001 . . . . .ohm*
ohm
1 statohm   = 8.987 551 79 × 10^{11} . . . . .ohms

(5) *ELECTRIC CAPACITANCE*

Electric capacitance has the dimensions electric charge per electric

$$\text{electric capacitance} = \frac{\text{electric charge}}{\text{electric potential}} = \frac{t^4 I^2}{m l^2}$$

potential. The farad is the SI unit and is equal to 1 coulomb per volt. In the SI base units the farad may be expressed as:

$$\frac{\text{coulomb}}{\text{volt}} = \text{ampere-second} \times \frac{\text{second}^3\text{-ampere}}{\text{kilogram-meter}^2}$$

$$= \frac{\text{second}^4\text{-ampere}^2}{\text{kilogram-meter}^2} = \text{farad}$$

The abfarad is equal to 1 abcoulomb per abvolt and the statfarad equals 1 statcoulomb per statvolt. Units in the tables are:

1 statfarad =   1.112 650 06 × 10^{-12} . . .farad
farad
1 abfarad   = 1 000 000 000 . . . . .farads*

(6) *ELECTRIC INDUCTANCE*

Electric inductance has the dimensions electric potential per electric

$$\text{electric inductance} = \frac{\text{electric potential}}{\text{electric current/time}}$$

$$= \frac{m l^2}{t^3 I} \times \frac{1}{I/t} = \frac{m l^2}{t^2 I^2}$$

current per time. This may be expressed as electric potential times time per electric current. The SI unit is the henry, which is equal to 1 volt-second per ampere. The henry may also be expressed as follows in base units:

$$\frac{\text{volt-second}}{\text{ampere}} = \frac{\text{kilogram-meter}^2}{\text{second}^3\text{-ampere}} \times \text{second} \times \frac{1}{\text{ampere}}$$

$$= \frac{\text{kilogram-meter}^2}{\text{second}^2\text{-ampere}^2} = \text{henry}$$

The abhenry equals 1 abvolt-second per abampere, and the stathenry is equal to 1 statvolt-second per statampere. Units in the tables are:

1 abhenry   = 0.000 000 001 . . . . .henry*
henry
1 stathenry   = 8.987 551 79 $\times$ 10^{11} . . . . .henries

(7) *ELECTRIC CONDUCTANCE*

Electric conductance has the dimensions electric current per electric

$$\text{electric conductance} = \frac{\text{electric current}}{\text{electric potential}} = \frac{t^3 I^2}{ml^2}$$

potential. The SI unit is the siemens and was formerly called the mho to illustrate the fact that the unit is the reciprocal of the ohm. The siemens is equal to 1 ampere per volt, which may also be expressed as follows:

$$\frac{\text{ampere}}{\text{volt}} = \frac{\text{second}^3\text{-ampere}}{\text{kilogram-meter}^2} \times \text{ampere} = \frac{\text{second}^3\text{-ampere}^2}{\text{kilogram-meter}^2} = \text{siemens}$$

The absiemens equals 1 abampere per abvolt and the statsiemens equals 1 statampere per statvolt. Units in the tables are the following:

1 statsiemens =   1.112 650 06 $\times$ 10^{-12} . .siemens
siemens
1 absiemens   = 1 000 000 000 . . . . .siemens*

(8) *MAGNETIC FLUX*

Magnetic flux has the dimensions: electric potential times time. The SI

$$\text{magnetic flux} = \text{electric potential} \times \text{time}$$

$$= \frac{m l^2}{t^2 I}$$

unit is the weber and it is equal to 1 volt-second. The weber may also be expressed as follows:

$$\text{volt-second} = \frac{\text{kilogram-meter}^2}{\text{second}^3\text{-ampere}} \times \text{second}$$

$$= \frac{\text{kilogram-meter}^2}{\text{second}^2\text{-ampere}} = \text{weber}$$

The tables include only 2 metric units, as is true of the tables of the other categories of units of magnetism. Electrostatic units have no special names and are excluded as not being very important. The maxwell is the electromagnetic unit and it equals 1 abvolt-second. The weber and the maxwell are the only units in the brief tables, with the following relationship:

$$1 \text{ maxwell} = 0.000\ 000\ 01 \quad \text{weber*}$$

(9) *MAGNETIC FLUX DENSITY, MAGNETIC INDUCTION*

Magnetic flux density or magnetic induction has the dimensions mag-

$$\text{magnetic induction} = \frac{\text{magnetic flux}}{\text{area}}$$

$$= \frac{m}{t^2 I}$$

netic flux per area. The tesla is the SI unit and is equal to 1 weber per square meter. The tesla may be expressed in the base units as follows:

$$\frac{\text{weber}}{\text{meter}^2} = \frac{\text{kilogram-meter}^2}{\text{second}^2\text{-ampere}} \times \frac{1}{\text{meter}^2} = \frac{\text{kilogram}}{\text{second}^2\text{-ampere}} = \text{tesla}$$

The gauss is the electromagnetic unit and is equal to 1 maxwell per square centimeter. The tesla and the gauss are the only units in the tables and have the following relationship:

$$1 \text{ gauss } = 0.000 \ 1 \quad \text{tesla*}$$

(10) *MAGNETIC FIELD STRENGTH*

Magnetic field strength has the dimensions electric current per length.

$$\text{magnetic field strength } = \frac{\text{electric current}}{\text{length}} = \frac{I}{l}$$

The SI unit is the ampere per meter and has no special name. It is thus expressed in base units.

The oersted is the electromagnetic unit. It is equal to $1/4\pi$ abampere-turn per centimeter. The ampere per meter and the oersted have the following relationship in the tables:

1 oersted corresponds to   1 000/4π amperes per meter*
1 oersted corresponds to 79.577 471 55   amperes per meter

(11) *MAGNETOMOTIVE FORCE, MAGNETIC POTENTIAL DIFFERENCE*

The ampere is the SI unit of magnetomotive force or magnetic potential difference. By convention the ampere is dimensionally independent. The ampere is defined as a constant current which produces a force between two conductors, and this is magnetomotive force.

The gilbert is the electromagnetic unit and is equal to $1/4\pi$ abampere-turn. The ampere and the gilbert are related as follows in the tables:

1 gilbert corresponds to 1/0.4π ampere*
1 gilbert corresponds to 0.795 774 715 5   ampere

17. ATOMIC ENERGY UNITS

Atomic energy units are energy units of a special nature. They cover a very broad numerical range from the electronvolt up to the energy equivalent of a kilogram mass in this book.

The first unit in the tables is the electronvolt, symbol eV, which is defined in the section "The International System of Units (SI)." One electronvolt equals 1 volt times the elementary charge, which is the charge of an electron. The following analysis shows that electric potential times electric charge has the dimensions of energy:

$$\text{volt-coulomb} = \frac{\text{kilogram-meter}^2}{\text{second}^3\text{-ampere}} \times \text{ampere-second}$$

$$= \frac{\text{kilogram-meter}^2}{\text{second}^2} = \text{joule}$$

Thus, 1 volt-coulomb equals 1 joule, the SI unit of energy. The elementary charge, symbol e, is from Cohen and Taylor (ref 2):

$$e = 1.602\ 189\ 2 \times 10^{-19}\quad \text{coulomb}$$

One electronvolt is equal to 1 volt times e, or $1.602\ 189\ 2 \times 10^{-19}$ joule.

The next two units in the tables are the kiloelectronvolt, symbol keV, which equals 1 000 eV, and the megaelectronvolt, symbol MeV, which equals 1 000 000 eV.

The unified atomic mass unit, symbol u, is defined in the section "The International System of Units (SI)." The energy equivalent of u is the fourth unit in the tables. The value of u is from Cohen and Taylor (ref 2):

$$u = 1.660\ 565\ 5 \times 10^{-27}\quad \text{kilogram}$$

The famous Einstein equation, $E = mc^2$, is used to calculate the energy equivalent of a unit of mass. If mass m is in kilograms and c, the speed of light in a vacuum, is in meters per second, the energy E is in joules. As noted previously, the value of c is from Cohen and Taylor (ref 2) and is repeated here for convenience:

$$c = 2.997\ 924\ 58 \times 10^8\quad \text{meters/second}$$

The following calculation yields the energy equivalent of u:

$$E = 1.660\ 565\ 5 \times 10^{-27}\ \text{kilogram}\ (2.997\ 924\ 58 \times 10^8\ \text{meters/second})^2$$

$$E = 1.492\ 441\ 8 \times 10^{-10}\ \text{joule}$$

The energy equivalent of a unit of mass is designated in this book by following the mass unit with the word "energy" in parentheses.

The joule, foot pound-force, and horsepower-hour are included to compare atomic energy units to units in everyday use. The last three units in the tables are the energy equivalents of the gram, pound, and kilogram. Einstein's equation is used to calculate the energies in joules.

Following is the list of units presented in the tables with all units related to the joule:

1 electrovolt	=	$1.602\ 189\ 2 \times 10^{-19}$......joule
1 kiloelectron-volt	=	$1.602\ 189\ 2 \times 10^{-16}$......joule
1 megaelectron-volt	=	$1.602\ 189\ 2 \times 10^{-13}$......joule
1 unified atomic mass unit (energy)	=	$1.492\ 441\ 8 \times 10^{-10}$......joule
joule		
1 foot pound-force	=	$1.355\ 817\ 948\ 331\ 400\ 4$..joules*
1 horsepower-hour	=	$2\ 684\ 519.537\ 696\ 172\ 792$.......joules*
1 gram (energy)	=	$8.987\ 551\ 79 \times 10^{13}$.....joules
1 pound (energy)	=	$4.076\ 684\ 92 \times 10^{16}$.....joules
1 kilogram (energy)	=	$8.987\ 551\ 79 \times 10^{16}$.....joules

Suggestions for Easing the Conversion to SI

When we say that a man is "six-foot-seven," we convey a clear picture of a very tall man. We know that there are 12 inches in a foot. Thus we know that he is over six and one-half feet tall. It is not very likely that we will convert the 7 inches to a decimal value and say that he is about 6.58 feet tall. In this situation we are accustomed to the use of mixed units.

In another situation we may say that a piece of wood is three and three-eighths inches wide. If we write this as a decimal value, 3.375, we appear to be saying that we know the width to the thousandth of an inch. This is very unlikely unless we used a high-precision instrument for the measurement. We would usually write this width as 3-3/8 inches and thus deal with fractions.

In many situations we are accustomed to dealing with decimals. Cash registers automatically ring-up sales in dollars and cents. We would not say we have 4 dollars, 5 quarters, and 3 dimes; we would say we have $5.55. Generally we prefer to deal with decimals because they are easier to handle than mixed units and fractions.

The International System of Units (SI) and other metric units are for the most part based upon the decimal system. Time and angular measure are exceptions.

Compare the following relationships involving units of length and mass:

1 meter = 100 centimeters = 1 000 millimeters; 1 kilometer = 1 000 meters
1 yard = 3 feet = 36 inches; 1 mile = 1 760 yards

1 kilogram = 1 000 grams = 1 000 000 milligrams
1 pound = 16 ounces = 7 000 grains

The metric relationships are obviously easier to learn, remember, and understand. The prefixes are the same for length and mass and are all powers of 10. On the other hand, we must learn each relationship in customary units individually; we must memorize the 3, 36, 1 760, 16, and 7 000. But by now we have them memorized and have to learn the metric relationships, that is, we have to learn to sense the pictures conveyed by a meter, kilometer, kilogram, etc.

41

We will have to make a considerable effort to learn to use SI and relate to its units. But after we have accomplished this feat, we will appreciate SI. There are many features about SI that make it much easier to use than the customary system. Here are some of them:

1. There are fewer units to learn. Most categories of measurement have a primary SI unit and a few useful multiples of this unit.

2. Most of the categories are linked together in a coherent system. The meter, kilogram, and second are the foundation for the SI units of area, force, pressure, energy, power, etc.

3. There is a set of short prefixes that apply to the fundamental units in all categories and always have the same meanings.

4. These prefixes may be substituted for numbers, which makes it easier to express values. For example, it is easier to write 2.5 micrograms rather than the equivalent 0.000 002 5 gram.

5. As mentioned previously, SI is mainly a decimal system. Mixed units are avoided except for time and angular measure and fractions are rarely necessary.

The purpose of this section is to make learning SI easier. Most of the units recommended for use are SI units and their multiples. Some metric units are recommended that are not part of SI but nevertheless are accepted for use with SI, notably the liter, hectare, and metric ton. Simplified conversion factors with few significant figures are presented, and asterisks are not used to indicate exact relationships; some require many significant figures to become exact.

Many multiples of SI units are recommended, rather than the base units, because they are numerically more compatible with the other unit in the relationship. The kilometer is related to the mile because 1 mile equals 1.609 kilometers, which is more compatible than the equivalent 1 609 meters. Some metric units which are not part of SI are provided with suggested SI replacements, such as the atmosphere, a non-SI unit of pressure, which equals about 101 kilopascals.

Frequent reference is made to the tables. These are the tables which are the major part of this book, the tables of usually 10-digit conversion factors.

Categories are listed in this section in the following order:

1. Length   6. Temperature
2. Area   7. Velocity
3. Volume   8. Flow
4. Mass   9. Density and Concentration
5. Time   10. Force

1. LENGTH

The tables include 7 metric units of length. The ones that are recommended for most practical applications are the millimeter, centimeter, meter, and kilometer, all SI units. The other 3 units are the angstrom, micrometer, and nautical mile. The angstrom will eventually be replaced because it is not an SI unit and it is close to the SI nanometer in size. The micrometer is the correct name of the unit often called the micron; it is useful for microscopic work. The nautical mile is not an SI unit and is best reserved for nautical purposes.

There are 9 customary units but only 4 are in general use: the inch, foot, yard, and mile. The other 5 units, which are the link, rod, chain, survey foot, and survey mile are surveyor's units.

The most important relationships are:

```
1 centimeter =       10. . . . . . . . . . . . . . millimeters
1 meter      =      100. . . . . . . . . . . . . centimeters
1 kilometer  = 1  000. . . . . . . . . . . . . . meters
1 inch       =      25.4 . . . . . . . . . . . . millimeters
             =       2.54 . . . . . . . . . . . centimeters
1 foot       =      30.5 . . . . . . . . . . . centimeters = 0.305 meter
1 yard       =       0.914. . . . . . . . . . . meter
1 mile       =       1.609. . . . . . . . . . . kilometers
```

A six-footer is about 183 centimeters, or 1.83 meters, tall. A 100-yard football field is 91.4 meters long; conversely, a field 100 meters long equals about 109 yards. A distance of 5 miles is equivalent to about 8 kilometers.

2. AREA

There are 6 metric units in the tables and 4 are recommended: the square centimeter, square meter, hectare, and square kilometer. All but the hectare are SI units. The hectare is a convenient name for the equivalent

SI unit, the square hectometer. Shorter names are often adopted for general use because of this convenience. The square millimeter is an SI unit useful for very small areas. The are is not an SI unit and may be confused with the verb or the acre.

There are 10 customary units but only 5 are popular. These are the square inch, square foot, square yard, acre, and square mile. The other 5 units are the squares of the surveyor's units of length. The acre is a surveyor's unit that has become popular.

The most important relationships are:

1 square meter	=	10 000	. . . . square centimeters
1 hectare (square hectometer)	=	10 000	. . . . square meters
1 square kilometer	=	100	. . . . hectares (square hectometers)
1 square inch	=	6.45	. . . . . square centimeters
1 square foot	=	929	. . . . square centimeters
	=	0.092 9	. . . square meter
1 square yard	=	0.836	. . . . square meter
1 acre	=	0.405	. . . . hectare = 4 050 sq meters
1 square mile	=	2.59	. . . . . square kilometers

An area of 1 000 square feet is about 93 square meters. An area of 100 square yards is about 84 square meters. A 40-acre lot equals almost 16.2 hectares. A geographical area of 1 000 square miles equals about 2 590 square kilometers.

3. VOLUME

The tables list 6 metric units of volume that are actually 3 pairs of equivalent units. The most useful units are the milliliter and liter, which are not SI units but which are nevertheless widely used and acceptable. Some may prefer to use the cubic centimeter, which is the SI equivalent of the milliliter. Probably fewer individuals will prefer to use the SI cubic decimeter instead of the liter. Again, short names are convenient, particularly in the case of the well-established liter. The other units are the kiloliter and its SI equivalent, the cubic meter, but this volume is usually too large for most practical applications.

There are 15 customary units in the tables, 8 of them called liquid measure units and 4 dry measure units. The units that are probably used most often are the cubic inch, cubic foot, fluid ounce, liquid pint, liquid quart, gallon, dry quart, and bushel. Of the remaining 7 units, the minim and fluid dram are quite small and the cubic yard is too large for much practical use. The gill, dry pint, and peck are used but not as frequently as the 8 listed. The petroleum barrel is included because of the increased public interest in petroleum.

Old habits can and should be broken, and the 2 metric units should replace the 15 customary units. Is it really necessary to maintain a set of volume units for special use as liquid or dry measure?

The most important relationships are:

1 liter	=	1 000.	milliliters
1 cubic meter	=	1 000.	liters
1 cubic inch	=	16.4.	milliliters
1 cubic foot	=	28.3.	liters
1 fluid ounce	=	29.6.	milliliters
1 liquid pint	=	0.473	liter = 473 milliliters
1 liquid quart	=	0.946	liter = 946 milliliters
1 gallon	=	3.79.	liters
1 dry quart	=	1.10.	liters
1 bushel	=	35.2.	liters

A liter of milk is equivalent to a little more than a quart, 1.8 ounces more. Ten gallons of gasoline equal about 38 liters. At 60 cents per gallon, gasoline costs about 16 cents per liter. A liter of strawberries has about 9% fewer strawberries than a dry quart. A fifth of whiskey, that is, one-fifth of a gallon, equals about 757 milliliters, about 3/4 of a liter.

4. MASS

Weight is the synonym generally used for mass and it is likely to remain so when used for nontechnical purposes.

All 4 metric units in the tables have practical use, and all but the metric ton are SI units. The metric ton is widely used and is acceptable for use with SI, but the proper SI name is megagram.

Of the 11 customary units, only 3 are popular: the avoirdupois ounce, avoirdupois pound, and short ton. In fact, it is not necessary to use the adjective "avoirdupois" in general use, but in tables that include apothecaries and troy units, the adjective is advisable. The other 8 units are used for technical and scientific purposes and are the grain, apothecaries scruple and dram, pennyweight, apothecaries or troy ounce and pound, and the avoirdupois dram and long ton.

The most important relationships are:

1 gram	= 1 000	milligrams
1 kilogram	= 1 000	grams
1 metric ton	= 1 000	kilograms
1 avoirdupois ounce	= 28.35	grams
1 avoirdupois pound	= 453.6	grams = 0.453 6 kilogram
1 short ton	= 0.907	metric ton = 907 kilograms

The milligram is a small unit, but it is widely used in medicinal drugs. A 100-pound individual weighs, or has a mass of, about 45.4 kilograms; 150 pounds equal 68.0 kilograms, and a 250-pound football player weighs about 113 kilograms. A 100-kilogram person is a heavyweight at a little over 220 pounds. A metric ton of coal is about 10% more coal than a short ton.

It is helpful to remember that 1 kilogram equals about 2.20 avoirdupois pounds, and 1 metric ton equals about 1.10 short tons.

5. TIME

The second is the SI and the customary unit of time. The other units in customary use are acceptable for use with SI and include the minute, hour, day, week, etc. There is no move under way to establish a system of units using SI prefixes such as the kilosecond and megasecond to replace the customary units. However, scientists and engineers are using smaller units including the millisecond, microsecond, and nanosecond.

We will continue to use mixed units such as 3 hours, 15 minutes, and 6 seconds. Stopwatches tell time in decimal parts of a second, but difficulties remain in converting other units to the decimal system. Is it easier to relate to 3.2517 hours instead of 3 hours, 15 minutes, and 6 seconds? Probably not.

6. TEMPERATURE

The Celsius scale is an old standby in the scientific and technical communities, and is not completely strange to the public. Celsius is the correct name of the scale formerly called centigrade, although some individuals are still using the latter name.

The basic relationship between the Celsius and Fahrenheit scales is that an interval of 5°C equals an interval of 9°F. The Celsius scale is a part of SI, although the SI unit of thermodynamic temperature is the kelvin. The unit "degree Celsius" is equal to the unit "kelvin."

The public will have to become more familiar with Celsius temperatures that equal key Fahrenheit temperatures. Following are a few examples:

freezing point of water	=	32°F = 0°C, exactly
boiling point of water	=	212°F = 100°C, exactly
normal human body temperature	=	98.6°F = 37°C, exactly
very hot afternoon	=	100°F = 37.8°C
very cold afternoon	=	0°F = -17.8°C
comfortable	=	68 to 77°F = 20 to 25°C, exactly

Table 1 in the temperature tables is a helpful guide because it presents exact relationships between the 2 scales from -454 to 1 112°F, or from -270 to 600°C. The other 4 tables are also presented to simplify conversions between the scales.

7. VELOCITY

The tables contain 6 metric units of velocity. The SI unit is the meter per second, which is a useful unit, but the kilometer per hour will be the metric unit most widely used by the public. The knot equals 1 nautical mile per hour and is widely used in such fields as navigation and meteorology; it is familiar to much of the public. The other 3 units are less useful and are the centimeter per second, meter per minute, and kilometer per minute.

The mile per hour is the most popular of the 5 customary units in the tables. The other 4 units are used to a lesser degree by the public but are included in the relationships presented below.

The most important relationships are:

1 knot	=	1.852.	kilometers per hour
	=	0.514.	meter per second
1 meter per second	=	3.6.	kilometers per hour
1 mile per hour	=	1.609	kilometers per hour
1 foot per minute	=	0.005 08	meter per second
1 foot per second	=	0.305	meter per second
1 mile per minute	=	96.6.	kilometers per hour

1 mile per second   = 5 790. . . . . kilometers per hour

The recommended limit of 55 miles per hour equals about 88.5 kilometers per hour, and 100 miles per hour sound much faster at 161 kilometers per hour. Perhaps the numerical increase that occurs when kilometers per hour replace miles per hour will tend to reduce highway speeds.

8. FLOW

There are only 5 metric units of flow in the tables and each one is given with 2 volume units. The SI unit is the cubic meter per second, which is too high a flow rate for most practical purposes. The liter per minute and liter per second should replace the customary units. The cubic decimeter may be preferred by some in place of the liter. The milliliter per second is used less than the other units.

There are 6 customary units and 4 are widely used: the gallon per minute, gallon per second, cubic foot per minute, and cubic foot per second. The petroleum barrel per hour and the cubic yard per minute are generally limited to technical use.

The most important relationships are:

1 cubic meter per second   = 100 0 . . . . liters per second
1 liter per second   =   60 . . . . liters per minute
1 gallon per minute   =   3.79 . . . . liters per minute
1 cubic foot per minute   =   28.3. . . . . liters per minute
1 gallon per second   =   3.79 . . . . liters per second
1 cubic foot per second   =   28.3 . . . . liters per second

9. DENSITY AND CONCENTRATION

There are only 3 numerically different metric units in the tables. The kilogram per cubic meter is the SI unit and it numerically equals the gram per liter, which is probably the most widely used unit. Both are in the tables grouped together. The other units are not used as often and are the gram per cubic meter or kiloliter and the gram per cubic centimeter or milliliter.

Of the 12 customary units in the tables, the pound per cubic foot and the pound per gallon are the most widely used. The other units are generally useful for scientific and technical purposes, and are too numerous to list here.

The most important relationships are:

1 pound per cubic foot =   16.0. . . . . grams per liter
1 pound per gallon   =   120 . . . . grams per liter

10. FORCE

There is a great deal of confusion about force and its relationship to mass and weight. Force equals mass times acceleration. Weight is a common synonym for mass, but it is really force. Unfortunately, many units of force have the same names as the units of mass from which they are derived. In this book, such units of force are clearly identified as units of force.

The SI unit of force is the newton and it is strongly recommended that the newton, and the related kilonewton, should replace all other units of force, both metric and customary. This would serve to eliminate the state of confusion.

Following is a list of the SI relationship and all of the non-SI units of force in the tables with the recommended conversion factors:

1 kilonewton   = 1 000. newtons
1 dyne   =   0.000 01 . . . . newton
1 gram-force   =   0.009 81 . . . . newton
1 kilogram-force =   9.81. . . . . newtons
1 metric ton-force=   9.81. . . . . kilonewtons
1 poundal   =   0.138 . . . . newton
1 pound-force   =   4.45. . . . . newtons
1 short ton-force =   8.90. . . . . kilonewtons

11. ENERGY

There are 13 metric units of energy in the tables. Included are the primary SI unit, which is the joule, plus the microjoule, kilojoule, and megajoule. The reason that the 3 related units are included is to show how they may be used together with the joule to replace the other 9 metric units as well as the 4 customary units in the tables.

Mass-related units should certainly be replaced; these are the gram-force centimeter, kilogram-force meter, and the foot pound-force. There are 5 energy units that are cumbersome compounds of power and time units that could very conveniently be replaced by SI units; these are the watt-hour, kilowatt-hour, and the metric, electric, and customary horsepower-hour units. The erg should be replaced by the microjoule, the foot poundal by the joule, and the Btu by the kilojoule.

The calorie and kilocalorie have exact relationships with SI units, but it would be advantageous to replace them with SI units to keep the number of energy units to a minimum. The kilocalorie is sometimes known as the large calorie and is used to evaluate the energy content of foods.

Following is a list of the SI relationships and all of the non-SI units of energy in the tables, both metric and customary, with the recommended conversion factors:

1 joule	=	1 000 000	microjoules
1 kilojoule	=	1 000	joules
1 megajoule	=	1 000 000	joules
1 erg	=	0.1	microjoule
1 gram-force centimeter	=	98.1	microjoules
1 calorie	=	4.184	joules
1 kilogram-force meter	=	9.81	joules
1 watt-hour	=	3.6	kilojoules
1 kilocalorie	=	4.184	kilojoules
1 metric horsepower-hour	=	2.648	megajoules
1 electric horsepower- hour	=	2.686	megajoules
1 kilowatt-hour	=	3.6	megajoules
1 foot poundal	=	0.042 1	joule
1 foot pound-force	=	1.356	joules
1 Btu	=	1.054	kilojoules
1 horsepower-hour	=	2.685	megajoules

There is great international concern with energy sources and consumption. It would be most advantageous if the units used to measure energy were limited to very few SI units. Foods could be evaluated in kilojoules instead of kilocalories. The megajoule should replace the kilowatt-hour on electricity bills. The heating value of a fuel such as coal should be given in kilojoules per kilogram or in megajoules per metric ton; presently heating values are given in Btu per pound or short ton.

12. POWER

There are 11 metric units of power in the tables, including the primary SI unit, which is the watt, and the related microwatt, kilowatt, and megawatt. Three of these units, excluding the megawatt, should replace the other 7 metric units as well as the 6 customary units in the tables.

Fortunately, the watt and kilowatt are popular units. For some individuals these units have an electrical connotation. This need not be the case. These SI units may be conveniently used to rate engines, as one example, replacing the customary horsepower. Many electrical appliances are presently rated in watts and kilowatts. The megawatt is included because it too is a popular unit used to rate the large power plants, particularly the new, large fossil-fuel and nuclear plants.

Of the many units to be replaced, 4 are mass-related and 3 are the closely related horsepower units. The remaining 6 units are all rather cumbersome units of energy per time.

Following is a list of the SI relationships and all of the non-SI units of power in the tables, both metric and customary, with the recommended conversion factors:

1 watt	=	1 000 000 . . . . . . .microwatts	
1 kilowatt	=	1 000 . . . . . . .watts	
1 megawatt	=	1 000 000 . . . . . . .watts	
1 erg per second	=	0.1 . . . . . .microwatt	
1 gram-force centi-meter/second	=	98.1 . . . . . .microwatts	
1 kilogram-force meter/minute	=	0.163 4. . .watt	
1 calorie per second	=	4.184 . . . .watts	
1 kilocalorie per minute	=	69.7 . . . . . .watts	
1 metric horsepower	=	735.5 . . . . . .watts	= 0.735 5 kilowatt
1 electric horse-power	=	746 . . . . . . .watts	= 0.746 kilowatt
1 foot pound-force per minute	=	0.022 6. . .watt	
1 Btu per hour	=	0.293 . . . .watt	
1 foot pound-force per second	=	1.356 . . . .watts	
1 Btu per minute	=	17.57 . . . . .watts	
1 horsepower	=	745.7 . . . . . .watts	= 0.745 7 kilowatt
1 Btu per second	=	1.054 . . . .kilowatts	

Some individuals may prefer to measure power in kilowatts and energy in kilowatt-hours. This similarity in terms is somewhat like the confusing mass and force units that share names. Another approach without confusion would be to use a unit such as the kilojoule for energy and the kilowatt for power. Power is energy per time, and the following list is presented to encourage the use of the SI units for both energy and power:

1 watt	=	1	joule per second
	=	3 600	joules per hour
	=	3.6	kilojoules per hour
1 kilowatt	=	1	kilojoule per second
	=	3 600	kilojoules per hour
	=	3.6	megajoules per hour
1 megawatt	=	1	megajoule per second
	=	3 600	megajoules per hour

All of the above relationships are exact, easy to understand, and all avoid the confusion of using similar names and the inconvenience of using long names like the kilowatt-hour.

13. PRESSURE

There are 11 metric units of pressure in the tables, which include the primary SI unit, called the pascal, and the kilopascal. These 2 units should replace the other metric units as well as the 3 customary units and the 4 units based on columns of water and mercury that are in the tables. In brief, 2 SI units should conveniently replace 16 non-SI units.

The pascal is the conveniently short name given recently to the newton per square meter. It is not well known to the public. The most popular units are the inch and millimeter of mercury and the pound-force per square inch. The atmosphere is widely used in science and the millibar is important in meteorology. The inch and foot of water are useful engineering units.

Many of the units are cumbersome compound units, including 7 mass-related force units per area and 2 true force units per area, which are the dyne and the newton per square centimeter. Columns of water and mercury are limited in accuracy by the measured densities of these liquids, which change with temperature; they are easily replaced by the SI units.

Four units have simple, exact relationships to SI units: the dyne per square centimeter equals 0.1 pascal, the millibar is equal to 0.1 kilopascal, the newton per square centimeter equals 10 kilopascals, and the bar equals 100 kilopascals. A fifth unit, the atmosphere, is defined as equal to exactly 101.325 kilopascals.

Following is a list of the SI relationship and all of the non-SI units of pressure in the tables with the recommended conversion factors:

1 kilopascal	=	1 000.	pascals
1 dyne per square centimeter	=	0.1.	pascal
1 kilogram-force/ square meter	=	9.81.	pascals
1 gram-force/ square centi- meter	=	98.1.	pascals
1 millibar	=	100.	pascals = 0.1 kilopascal
1 metric ton- force/sq meter	=	9.81.	kilopascals
1 newton per square centi- meter	=	10.	kilopascals
1 kilogram-force/ sq centimeter	=	98.1.	kilopascals
1 bar	=	100.	kilopascals
1 atmosphere	=	101.325.	kilopascals
1 pound-force per square foot	=	47.9.	pascals
1 pound-force per square inch	=	6.89.	kilopascals
1 short ton-force/ square foot	=	95.8.	kilopascals
1 millimeter of mercury	=	133.3.	pascals
1 inch of water	=	249.	pascals = 0.249 kilopascal
1 foot of water	=	2.99.	kilopascals
1 inch of mercury	=	3.39.	kilopascals

In the near future we should be hearing that the barometric pressure is 101.3 kilopascals and rising instead of 29.92 inches of mercury and rising. Tires would be inflated to 179 kilopascals instead of 26 pounds-force per square inch.

14. ANGULAR MEASURE

The SI unit of angular measure is the radian. Relatively inexpensive small calculators are available with a π-key, which makes the radian a unit that could conveniently replace other units. The 4 most popular units of angular measure and their relationships to the radian are as follows:

$$
\begin{aligned}
\text{1 second} &= 2\pi/1\ 296\ 000 \ \dots \text{radian} \\
&= 0.000\ 004\ 848\ 1 \dots \text{radian} \\
\text{1 minute} &= 2\pi/21\ 600 \dots \text{radian} \\
&= 0.000\ 290\ 888\ 2 \dots \text{radian} \\
\text{1 degree} &= 2\pi/360 \dots \text{radian} \\
&= 0.017\ 453\ 292\ 5 \dots \text{radian} \\
\text{1 circle} &= 2\pi \dots \text{radians} \\
&= 6.283\ 185\ 307 \dots \text{radians}
\end{aligned}
$$

A wheel may revolve through 10 circles or revolutions, 45 degrees, 12 minutes, and 6.3 seconds. This expression in 4 mixed units could be reduced to radians as follows:

$$
\begin{aligned}
\text{6.3 seconds} &= 6.3(2\pi/1\ 296\ 000) &= 0.000\ 030\ 543\ 3 \ \text{radian} \\
\text{12 minutes} &= 12\ (2\pi/21\ 600) &= 0.003\ 490\ 658\ 5 \ \text{radian} \\
\text{45 degrees} &= 45\ (2\pi/360) &= 0.785\ 398\ 163\ 4 \ \text{radian} \\
\text{10 circles} &= 10\ (2\pi) &= 62.831\ 853\ 07 \ \ \text{radians}
\end{aligned}
$$

The sum of the 4 calculated values equals 63.620 772 44 radians. This sum is carried out to 10 significant figures because the second is such a small unit compared to the circle.

The radian offers the simplest expression of angular measure. Mixed units are cumbersome. Decimals may be used but the relationships are not multiples of 10. It is not obvious that 45.201 75 degrees are equal to 45 degrees, 12 minutes, and 6.3 seconds.

For convenience, the value of π is listed below:

$$\pi = 3.141\ 592\ 653\ 590$$

15. ELECTRICITY AND MAGNETISM

There are 11 major categories of measurement of electricity and magnetism in this book, and each has an SI unit. Fortunately, for those individuals familiar with electricity and magnetism, the SI units are those that are in everyday use.

The SI units of electricity are the following:

ampere for current	farad for capacitance
coulomb for charge	henry for inductance
volt for potential	siemens for conductance
ohm for resistance	

The siemens is the name adopted recently for the unit formerly known as the mho.

The SI units of magnetism are the following:

weber for magnetic flux
tesla for magnetic flux density
ampere per meter for magnetic field strength
ampere for magnetomotive force

There is no ampere-turn in SI; this unit appears in references as the basic unit of magnetomotive force. By definition, the ampere is the SI base unit for both electric current and for magnetomotive force.

For those in the sciences who prefer to use electromagnetic and electrostatic systems of units for electricity and magnetism, the relationship between these systems and SI is quite complicated. The matter is discussed in the chapter "Categories and Units of Measurement."

16. LIGHT

The candela is the SI unit of luminous intensity; it defines the source of light and replaces previously used units. The lumen is the SI unit of luminous flux; it defines the rate at which light falls upon a surface. There are no other important units of luminous intensity or luminous flux.

There are 2 other categories of measurement of light in this book. The candela per square meter is the SI unit of luminance, and it should replace the other 6 units in the tables with no difficulties. The most important relationships in luminance are:

1 millilambert	=	3.18	candelas per square meter
1 footlambert	=	3.43	candelas per square meter
1 candela per square foot	=	10.76	candelas per square meter
1 candela per square inch	=	1 550	candelas per square meter
1 lambert	=	3 183	candelas per square meter
1 stilb	=	10 000	candelas per square meter

The final category is called illuminance or illumination. The SI unit is the lux, which equals 1 lumen per square meter. There are only 3 other units in the tables and they are related to the lux as follows:

1 milliphot   =   10 . . . . lux
1 footcandle =   10.76 . . . . lux
1 phot   =   10 000 . . . . lux

17. ATOMIC ENERGY UNITS

Atomic energy units are not popular, everyday units and one may guess that they are of special interest to the author. Nevertheless, Einstein's equation that relates energy to mass is very well known to the public, and there is a general fascination with the enormous amount of energy available from small quantities of mass.

A 1 000-megawatt power plant produces energy at the rate of 1 000 megajoules per second, or 3 600 000 megajoules per hour, or 31 536 000 000 megajoules per year. One kilogram is equivalent to about 90 000 000 000 megajoules. Thus the energy equivalent of 1 kilogram equals the output of the 1 000-megawatt plant for a period of 2.85 years, or over 2 years and 10 months.

To revert to customary units, 1 pound is equivalent to about 15 190 000 000 horsepower-hours. This is enough energy to operate a 300-horsepower engine for 5 779 years.

There are no SI units of atomic energy as such. However, there are some suggestions in order here. The electronvolt is one word. The mega-electronvolt should be used instead of the million-electronvolt unit. Because the billion means different things to different people, the giga-electronvolt, symbol GeV, should replace the billion-electronvolt unit.

TABLES OF RECOMMENDED UNITS

The table below lists all of the categories of measurement covered in this chapter excluding atomic energy units, and presents the units recommended for general use. The SI units, including those with proper SI prefixes, are indicated by asterisks.

Category of Measurement	Recommended Units
length	millimeter*, centimeter*, meter*, kilometer*
area	square centimeter*, square meter*, hectare, square kilometer*
volume	milliliter, liter
mass	milligram*, gram*, kilogram*, metric ton
time	second*, minute, hour, day, week, months, years
temperature	degree Celsius*
velocity	meter per second*, kilometer per hour
flow	liter per minute, liter per second
density and concentration	gram per liter
force	newton*, kilonewton*
energy	microjoule*, joule*, kilojoule*, megajoule*
power	microwatt*, watt*, kilowatt*, megawatt*
pressure	pascal*, kilopascal*
angular measure	radian*
electric current	ampere*
electric charge	coulomb*
electric potential	volt*
electric resistance	ohm*
electric capacitance	farad*
electric inductance	henry*
electric conductance	siemens*
magnetic flux	weber*
magnetic flux density	tesla*
magnetic field strength	ampere per meter*
magnetomotive force	ampere*
luminance	candela per square meter*
illuminance or illumination	lux*

Lists of Additional Units

Each category of measurement, except temperature, has a list of additional units following its tables. The primary purpose served by these lists is to provide a reasonably complete catalog of units that are less often used for general purposes than those in the tables.

The following units included in the list of additional units of length are used to illustrate reasons why units have been included:

1. The decimeter, hectometer, fathom, and furlong are examples of units in general use but less so than those in the tables.

2. The printer's point, the bolt of cloth, and the skein are examples of units which are important in special fields of interest.

3. The U.S. statute mile is included because it is a more definitive name of the unit known as the survey mile.

As in the tables, exact conversion factors are indicated by an asterisk following the relationship. Where the additional unit exactly equals one in the tables, the relationship is listed as follows:

1. U.S. statute mile = 1 survey mile* (in tables)

In all categories where applicable, factors are listed to convert all additional units to either the SI unit or to accepted multiples of the SI unit. The objective is to encourage use of SI and its multiples, which is a primary objective of the entire book.

Barbrow and Judson (ref 1) is an excellent source of units for the categories length, mass, area, and volume. Mechtly (ref 3) and ASTM (ref 6) provide units for many categories. Handbook (ref 8) includes by far the largest list of units of all the references consulted. In addition to these sources, several units were added by the author.

Where necessary, conversion factors are presented which serve to define the unit in terms of another customary unit. This factor is then followed by one to convert the unit to the SI unit or a multiple, as in the following example:

$$
\begin{aligned}
1 \text{ furlong} &= 660 \qquad \text{survey feet*} \\
&= 201.168 \text{ 4 meters}
\end{aligned}
$$

58

Inexact relationships are usually limited to only 7 significant figures. In some cases additional units which exactly equal units in the tables are not followed by a factor converting the unit to SI; substituting the name of the additional unit for the one in the tables and referring to the tables provides a complete set of conversion factors. In other situations where all additional units are converted to the SI unit, such as additional units of power, conversion factors are provided for all units so that there is a complete set of factors in one convenient place.

In four categories, AREA, LENGTH, MASS, and VOLUME, factors are provided to convert the additional units to SI units or other metric units. In many cases metric units are selected in lieu of the SI unit because these units are numerically closer to the size of the additional unit. For example, the measuring teaspoon, an additional unit of volume, is equivalent to almost 5 milliliters or 0.000 005 cubic meter. The SI unit of volume is the cubic meter, but the factor for converting to the milliliter is listed because the units are more similar numerically.

In four other categories, ENERGY, FORCE, POWER, and PRESSURE, factors are provided to convert all additional units to only SI units. Additional units of VELOCITY include factors to convert to both the SI unit, which is the meter per second, and to the widely used kilometer per hour.

Additional categories as well as additional units are presented for ELECTRICITY AND MAGNETISM and LIGHT. There are only 4 additional units listed for ANGULAR MEASURE, and only 5 for TIME. There are 3 additional units in ATOMIC ENERGY UNITS plus a table which lists both the masses and the equivalent energies of 3 sub-atomic particles: the electron, proton, and neutron. The rest masses and energies are taken from Cohen and Taylor (ref 2).

Throughout this book units are listed in ascending order and one group of exceptions. The additional units of ELECTRICITY AND MAGNETISM list the electromagnetic unit followed by the electrostatic unit as the first two units for the categories of electricity, and list the electromagnetic units as the first unit for the categories of magnetism.

A group called international units were excluded from the lists of additional units to avoid confusion with SI. These units have been replaced by SI units for all practical purposes. Their relationships to SI units are presented below as taken from Handbook (ref 8):

1 international ampere   = 0.999 835 . . . . ampere
1 international coulomb = 0.999 835 . . . . coulomb
1 international farad   = 0.999 505 . . . . farad

1 international henry	= 1.000 495	. . . . henries
1 international joule	= 1.000 165	. . . . joules
1 international mho	= 0.999 505	. . . . siemens
1 international ohm	= 1.000 495	. . . . ohms
1 international volt	= 1.000 330	. . . . volts
1 international watt	= 1.000 165	. . . . watts

Angular Measure

One SECOND is equal to:

$$0.016\ 666\ 666\ 67 \dots \dots \dots \dots \dots \text{minute}$$
$$0.000\ 308\ 641\ 975\ 3 \dots \dots \dots \dots \dots \text{grade}$$
$$0.000\ 277\ 777\ 777\ 8 \dots \dots \dots \dots \dots \text{degree}$$
$$0.000\ 004\ 848\ 136\ 811 \dots \dots \dots \dots \text{radian}$$
$$0.000\ 003\ 086\ 419\ 753 \dots \dots \dots \dots \text{quadrant}$$
$$7.716\ 049\ 383 \times 10^{-7} \dots \dots \dots \dots \text{circle}$$

One MINUTE is equal to:

$$60 \dots \dots \dots \dots \dots \dots \dots \dots \dots \dots \dots \text{seconds*}$$
$$0.018\ 518\ 518\ 52 \dots \dots \dots \dots \dots \text{grade}$$
$$0.016\ 666\ 666\ 67 \dots \dots \dots \dots \dots \text{degree}$$
$$0.000\ 290\ 888\ 208\ 7 \dots \dots \dots \dots \text{radian}$$
$$0.000\ 185\ 185\ 185\ 2 \dots \dots \dots \dots \text{quadrant}$$
$$0.000\ 046\ 296\ 296\ 30 \dots \dots \dots \dots \text{circle}$$

One GRADE is equal to:

$$3\ 240 \dots \dots \dots \dots \dots \dots \dots \dots \dots \text{seconds*}$$
$$54 \dots \dots \dots \dots \dots \dots \dots \dots \dots \dots \text{minutes*}$$
$$0.9 \dots \dots \dots \dots \dots \dots \dots \dots \dots \dots \text{degree*}$$
$$0.015\ 707\ 963\ 27 \dots \dots \dots \dots \dots \text{radian}$$
$$0.01 \dots \dots \dots \dots \dots \dots \dots \dots \dots \dots \text{quadrant*}$$
$$0.002\ 5 \dots \dots \dots \dots \dots \dots \dots \dots \dots \text{circle*}$$

One DEGREE is equal to:

$$3\ 600 \dots \dots \dots \dots \dots \dots \dots \dots \dots \text{seconds*}$$
$$60 \dots \dots \dots \dots \dots \dots \dots \dots \dots \dots \text{minutes*}$$
$$1.111\ 111\ 111 \dots \dots \dots \dots \dots \dots \text{grades}$$

61

```
0.017 453 292 52 . . . . . . . . . . . . . . . . . radian
0.011 111 111 11 . . . . . . . . . . . . . .quadrant

0.002 777 777 778 . . . . . . . . . . . . . . . . circle
```

One RADIAN is equal to:

```
206 264.806 2 . . . . . . . . . . . . . . . . . . . . . . . seconds
  3 437.746 771 . . . . . . . . . . . . . . . . . . . . . minutes
     63.661 977 24 . . . . . . . . . . . . . . . . . . . grades
     57.295 779 51 . . . . . . . . . . . . . . . . . . .degrees
      0.636 619 772 4 . . . . . . . . . . . . . . . .quadrant

      0.159 154 943 1 . . . . . . . . . . . . . . . . . . circle
```

One QUADRANT is equal to:

```
324 000 . . . . . . . . . . . . . . . . . . . . . . . . . . . .seconds*
  5 400 . . . . . . . . . . . . . . . . . . . . . . . . . . . . minutes*
    100 . . . . . . . . . . . . . . . . . . . . . . . . . . . .grades*
     90 . . . . . . . . . . . . . . . . . . . . . . . . . . . .degrees*
      1.570 796 327 . . . . . . . . . . . . . . . . . . . . radians

      0.25 . . . . . . . . . . . . . . . . . . . . . . . . . . circle*
```

One CIRCLE is equal to:

```
1 296 000 . . . . . . . . . . . . . . . . . . . . . . . . . . .seconds*
   21 600 . . . . . . . . . . . . . . . . . . . . . . . . . . . minutes*
      400 . . . . . . . . . . . . . . . . . . . . . . . . . . . .grades*
      360 . . . . . . . . . . . . . . . . . . . . . . . . . . . degrees*
        6.283 185 307 . . . . . . . . . . . . . . . . . . . radians

        4 . . . . . . . . . . . . . . . . . . . . . . . . . . quadrants*
```

ADDITIONAL UNITS OF ANGULAR MEASURE

```
1 centesimal minute =  0.01 . . . . . . . . . . . . . . . . . . . . . . . . grade*
1 sign              = 30 . . . . . . . . . . . . . . . . . . . . . . . . .degrees*
1 circumference     =  1 . . . . . . . . . . . . . . . . . . . . . circle* (in tables)
1 revolution        =  1 . . . . . . . . . . . . . . . . . . . . . circle* (in tables)
```

Area

One SQUARE MILLIMETER is equal to:

0.01 . . . . square centimeter*
0.001 550 003 100 . . . . square inch
0.000 024 710 439 30 . . . . square link
0.000 010 763 910 42 . . . . square foot
0.000 010 763 867 36 . . . . . . square survey foot

0.000 001 195 990 046 . . . . square yard
0.000 001 . . . . square meter*
3.953 670 289 $\times$ 10^{-8} . . . . square rod
0.000 000 01 . . . . are*
2.471 043 930 $\times$ 10^{-9} . . . . square chain

2.471 043 930 $\times$ 10^{-10} . . . . acre
0.000 000 000 1 . . . . hectare*
0.000 000 000 001 . . . . square kilometer*
3.861 021 585 $\times$ 10^{-13} . . . . square mile
3.861 006 141 $\times$ 10^{-13} . . . . . square survey mile

One SQUARE CENTIMETER is equal to:

100 . . . . square millimeters*
0.155 000 310 0 . . . . square inch
0.002 471 043 930 . . . . square link
0.001 076 391 042 . . . . square foot
0.001 076 386 736 . . . . square survey foot

0.000 119 599 004 6 . . . . square yard
0.000 1 . . . . square meter*
0.000 003 953 670 289 . . . . square rod
0.000 001 . . . . are*
2.471 043 930 $\times$ 10^{-7} . . . . square chain

2.471 043 930 $\times$ 10^{-8} . . . . acre

63

0.000 000 01. . . . . .hectare*
0.000 000 000 1 . . . . .square kilometer*
3.861 021 585 × 10⁻¹¹ . . . . square mile
3.861 006 141 × 10⁻¹¹ . . . . . square survey mile

One SQUARE INCH is equal to:

645.16. . . . . .square millimeters*
6.451 6 . . . . square centimeters*
0.015 942 187 02 . . . . square link
0.006 944 444 444 . . . . square foot
0.006 944 416 667 . . . . square survey foot

0.000 771 604 938 3. . . . . .square yard
0.000 645 16. . . . . square meter*
0.000 025 507 499 23 . . . . .square rod
0.000 006 451 6 . . . . .are*
0.000 001 594 218 702 . . . . square chain

1.594 218 702 × 10⁻⁷. . . . . .acre
0.000 000 064 516 . . . . .hectare*
0.000 000 000 645 16. . . . . . .square kilometer*
2.490 976 686 ×10⁻¹⁰ . . . . square mile
2.490 966 722 × 10⁻¹⁰ . . . . . square survey mile

One SQUARE LINK is equal to:

40 468.726 10 . . . . square millimeters
404.687 261 0. . . . . square centimeters
62.726 650 91. . . . . .square inches
0.435 601 742 4 . . . . square foot
0.435 6 . . . . square survey foot*

0.048 400 193 60 . . . . square yard
0.040 468 726 10 . . . . square meter
0.001 6 . . . . .square rod*
0.000 404 687 261 0. . . . . .are
0.000 1 . . . . square chain*

0.000 01 . . . . .acre*
0.000 004 046 872 610 . . . . .hectare
4.046 872 610 × 10⁻⁸ . . . . square kilometer

1.562 506 250 × 10⁻⁸ . . . . square mile
0.000 000 015 625 . . . . square survey mile*

One SQUARE FOOT is equal to:

92 903.04. . . . . .square millimeters*
929.030 4 . . . . square centimeters*
144. . . . . .square inches*
2.295 674 931 . . . . .square links
0.999 996 000 004 . . . . square survey foot*

0.111 111 111 1 . . . . .square yard
0.092 903 04. . . . . square meter*
0.003 673 079 890 . . . . .square rod
0.000 929 030 4 . . . . .are*
0.000 229 567 493 1. . . . . square chain

0.000 022 956 749 31. . . . . .acre
0.000 009 290 304 . . . . .hectare*
0.000 000 092 903 04. . . . . . .square kilometer*
3.587 006 428 × 10⁻⁸ . . . . square mile
3.586 992 080 × 10⁻⁸ . . . . . . square survey mile

One SQUARE SURVEY FOOT is equal to:

92 903.411 61 . . . . square millimeters
929.034 116 1. . . . . square centimeters
144.000 576 0. . . . . .square inches
2.295 684 114. . . . . .square links
1.000 004 000 . . . . square feet

0.111 111 555 6 . . . . .square yard
0.092 903 411 61 . . . . square meter
0.003 673 094 582 . . . . .square rod
0.000 929 034 116 1. . . . . .are
0.000 229 568 411 4. . . . . square chain

0.000 022 956 841 14. . . . . .acre
0.000 009 290 341 161 . . . . .hectare
9.290 341 161 × 10⁻⁸ . . . . square kilometer
3.587 020 776 × 10⁻⁸ . . . . square mile
3.587 006 428 × 10⁻⁸ . . . . . . square survey mile

One SQUARE YARD is equal to:

836 127.36. . . . . .square millimeters*
8 361.273 6 . . . . square centimeters*
1 296. . . . . .square inches*
20.661 074 38. . . . . .square links
9. . . . . square feet*

8.999 964 000 036 . . . . .square survey feet*
0.836 127 36. . . . . square meter*
0.033 057 719 01 . . . . .square rod
0.008 361 273 6 . . . . .are*
0.002 066 107 438 . . . . square chain

0.000 206 610 743 8. . . . . .acre
0.000 083 612 736 . . . . .hectare*
0.000 000 836 127 36. . . . . . .square kilometer*
3.228 305 785 × 10^{-7}. . . . . square mile
3.228 292 872 × 10^{-7}. . . . . . square survey mile

One SQUARE METER is equal to:

1 000 000. . . . . .square millimeters*
10 000. square centimeters*
1 550.003 100 . . . . .square inches
24.710 439 30. . . . . .square links
10.763 910 42. . . . . square feet

10.763 867 36. . . . . .square survey feet
1.195 990 046. . . . . square yards
0.039 536 702 89 . . . . .square rod
0.01. . . . . .are*
0.002 471 043 930 . . . . square chain

0.000 247 104 393 0. . . . . .acre
0.000 1 . . . . .hectare*
0.000 001 . . . . .square kilometer*
3.861 021 585 × 10^{-7}. . . . . square mile
3.861 006 141 × 10^{-7}. . . . . . square survey mile

One SQUARE ROD is equal to:

25 292 953.81. square millimeters

252 929.538 1 . . . . square centimeters
 39 204.156 82 . . . . .square inches
     625. . . . . .square links*
   272.251 089 0. . . . . square feet

   272.25. . . . . .square survey feet*
    30.250 121 00. . . . . square yards
    25.292 953 81. . . . . square meters
     0.252 929 538 1 . . . . .are
     0.062 5 . . . . square chain*

     0.006 25 . . . . acre*
     0.002 529 295 381 . . . . .hectare
     0.000 025 292 953 81 . . . . square kilometer
     0.000 009 765 664 063 . . . . .square mile
     0.000 009 765 625 . . . . square survey mile*

One ARE is equal to:

100 000 000. .square millimeters*
  1 000 000. . . . . square centimeters*
   155 000.310 0 .square inches
     2 471.043 930 . . . . .square links
     1 076.391 042 . . . . square feet

     1 076.386 736 . . . . .square survey feet
      119.599 004 6. . . . . square yards
      100. . . . . .square meters*
        3.953 670 289. . . . . square rods
        0.247 104 393 0 . . . . square chain

        0.024 710 439 30 . . . . .acre
        0.01. . . . . .hectare*
        0.000 1 . . . . .square kilometer*
        0.000 038 610 215 85. . . . . square mile
        0.000 038 610 061 41. . . . . . square survey mile

One SQUARE CHAIN is equal to:

404 687 261.0 . . . . square millimeters
  4 046 872.610. . . . . square centimeters
   627 266.509 1 .square inches
    10 000. . . . . .square links*

4 356.017 424 . . . . square feet

4 356. . . . . .square survey feet*
484.001 936 0. . . . . square yards
404.687 261 0. . . . . square meters
16. . . . . square rods*
4.046 872 610. ares

0.1 . . . . acre*
0.040 468 726 10 . . . . .hectare
0.000 404 687 261 0. . . . . square kilometer
0.000 156 250 625 0. . . . . square mile
0.000 156 25. . . . . square survey mile*

One ACRE is equal to:

4.046 872 610 × 10⁹ . . . . . . square millimeters
40 468 726.10. . . . . square centimeters
6 272 665.091. . . . . .square inches
100 000. . . . . .square links*
43 560.174 24 . . . . square feet

43 560. . . . . .square survey feet*
4 840.019 360 . . . . square yards
4 046.872 610 . . . . square meters
160. . . . . square rods*
40.468 726 10. . . . . .ares

10. . . . . .square chains*
0.404 687 261 0 . . . . .hectare
0.004 046 872 610 . . . . square kilometer
0.001 562 506 250 . . . . square mile
0.001 562 5. . . . . square survey mile*

One HECTARE is equal to:

10 000 000 000. .square millimeters*
100 000 000. . . . . square centimeters*
15 500 031.00. . . . . .square inches
247 104.393 0 . . . . .square links
107 639.104 2 . . . . square feet

```
  107 638.673 6 . . . . . . . . . . . . . . . . .square survey feet
   11 959.900 46 . . . . . . . . . . . . . . . . . . square yards
   10 000. . . . . . . . . . . . . . . . . . . . . . . .square meters*
      395.367 028 9. . . . . . . . . . . . . . . . square rods
      100. . . . . . . . . . . . . . . . . . . . . . . . . . . . ares*

       24.710 439 30. . . . . . . . . . . . . . . .square chains
        2.471 043 930 . . . . . . . . . . . . . . . . . . . acres
        0.01. . . . . . . . . . . . . . . . . . . .square kilometer*
        0.003 861 021 585 . . . . . . . . . . . . square mile
        0.003 861 006 141 . . . . . . . . square survey mile
```

One SQUARE KILOMETER is equal to:

```
1 000 000 000 000. . . . . . . . . . . . . . . . . . . . .square millimeters*
   10 000 000 000. . . . . . . . . . . . . . . . . . . square centimeters*
            1.550 003 100 × 10⁹ . . . . . . . . .square inches
      24 710 439.30. . . . . . . . . . . . . . . . . . . . .square links
      10 763 910.42. . . . . . . . . . . . . . . . . . . . . square feet

      10 763 867.36. . . . . . . . . . . . . . . . . . . .square survey feet
       1 195 990.046. . . . . . . . . . . . . . . . . . . . . square yards
       1 000 000. . . . . . . . . . . . . . . . . . . . . . .square meters*
          39 536.702 89 . . . . . . . . . . . . . . . . . . square rods
          10 000. . . . . . . . . . . . . . . . . . . . . . . . . . . ares*

       2 471.043 930 . . . . . . . . . . . . . . . . . .square chains
         247.104 393 0. . . . . . . . . . . . . . . . . . . . . acres
         100. . . . . . . . . . . . . . . . . . . . . . . . . . . hectares*
           0.386 102 158 5 . . . . . . . . . . . . . . square mile
           0.386 100 614 1 . . . . . . . . . . square survey mile
```

One SQUARE MILE is equal to:

```
2 589 988 110 336. . . . . . . . . . . . . . . . . . . . .square millimeters*
   25 899 881 103.36. . . . . . . . . . . . . . . . . . square centimeters*
    4 014 489 600. . . . . . . . . . . . . . . . . . . . . . .square inches*
       63 999 744.000 256 . . . . . . . . . . . . . . . . . .square links*
       27 878 400. . . . . . . . . . . . . . . . . . . . . . . . square feet*

       27 878 288.486 511 513 6 . . . . . . . . . .square survey feet*
        3 097 600. . . . . . . . . . . . . . . . . . . . . . . . square yards*
```

2 589 988.110 336 . . . . .square meters*
102 399.590 400 409 6 . . . . square rods*
25 899.881 103 36. ares*

6 399.974 400 025 6 . . . . square chains*
639.997 440 002 56 . . . . acres*
258.998 811 033 6 . . . . hectares*
2.589 988 110 336 . . . . square kilometers*
0.999 996 000 004 . . . . square survey mile*

One **SQUARE SURVEY MILE** is equal to:

2.589 998 470 $\times$ 10^{12} . . . . . square millimeters
2.589 998 470 $\times$ 10^{10} . . . . . . square centimeters
4.014 505 658 $\times$ 10^{9} . . . . .square inches
64 000 000. . . . . .square links*
27 878 511.51. . . . . square feet

27 878 400. . . . . .square survey feet*
3 097 612.390. . . . . square yards
2 589 998.470. . . . . square meters
102 400. . . . . square rods*
25 899.984 70 . . . . ares

6 400. . . . . .square chains*
640. . . . . acres*
258.999 847 0. . . . . hectares
2.589 998 470. . . . . .square kilometers
1.000 004 000. . . . . square miles

ADDITIONAL UNITS OF AREA

Metric

1 barn	=	1×10^{-24} . . . . . . .square centimeter*
1 circular millimeter	=	0.785 398 2 . . . . square millimeter
1 square decimeter	=	0.01 . . . . square meter*
1 centare	=	1 . . . . square meter* (in tables)
1 square dekameter	=	100 . . . . .square meters*
	=	1 . . . . .are* (in tables)
1 square hectometer	=	10 000 . . . . .square meters*
	=	1 . . . . .hectare* (in tables)

Customary

1 circular mil	=	0.785 398 2 . . . . .square mil
	=	7.853 982 $\times$ 10^{-7} . . . . square inch
	=	0.000 506 707 5 . . . . square millimeter
1 square mil	=	0.000 001. . . . . square inch*
	=	0.000 645 16 . . . . . . square millimeter*
1 circular inch	=	0.785 398 2. . . . . square inch
	=	5.067 075. . . . . square centimeters

1 Gunter's square link	=	1 . . . . square link* (in tables)
1 surveyor's square link	=	1 . . . . square link* (in tables)
1 engineer's square link	=	1 . . . . square foot* (in tables)
1 Ramden's square link	=	1 . . . . square foot* (in tables)

1 building square	=	100 . . . . square feet*
	=	9.290 304. . . . . .square meters*
1 square perch	=	1 . . . . .square rod* (in tables)
1 square pole	=	1 . . . . .square rod* (in tables)

1 Gunter's square chain	=	1 . . . . square chain* (in tables)
1 surveyor's square chain	=	1 . . . . square chain* (in tables)
1 engineer's square chain	=	10 000 . . . . square feet*
	=	929.030 4 . . . . .square meters*
1 Ramden's square chain	=	10 000 . . . . square feet*
	=	929.030 4 . . . . .square meters*

1 section of land	=	1 . . . . . . square survey mile* (in tables)
1 township	=	36 . . . . .square survey miles*
	=	93.239 94 . . . . .square kilometers

Atomic Energy Units

One ELECTRONVOLT is equal to:

$$0.001 \dots \text{kiloelectronvolt*}$$
$$0.000\ 001 \dots \text{megaelectronvolt*}$$
$$1.073\ 535\ 4 \times 10^{-9} \dots \text{unified atomic mass}$$
$$\text{unit (energy)}$$
$$1.602\ 189\ 2 \times 10^{-19} \dots \text{joule}$$
$$1.181\ 714\ 1 \times 10^{-19} \dots \text{foot pound-force}$$

$$5.968\ 253 \times 10^{-26} \dots \text{horsepower-hour}$$
$$1.782\ 675\ 9 \times 10^{-33} \dots \text{gram (energy)}$$
$$3.930\ 128 \times 10^{-36} \dots \text{pound (energy)}$$
$$1.782\ 675\ 9 \times 10^{-36} \dots \text{kilogram (energy)}$$

One KILOELECTRONVOLT is equal to:

$$1\ 000 \dots \text{electronvolts*}$$
$$0.001 \dots \text{megaelectronvolt*}$$
$$0.000\ 001\ 073\ 535\ 4 \dots \text{unified atomic mass}$$
$$\text{unit (energy)}$$
$$1.602\ 189\ 2 \times 10^{-16} \dots \text{joule}$$
$$1.181\ 714\ 1 \times 10^{-16} \dots \text{foot pound-force}$$

$$5.968\ 253 \times 10^{-23} \dots \text{horsepower-hour}$$
$$1.782\ 675\ 9 \times 10^{-30} \dots \text{gram (energy)}$$
$$3.930\ 128 \times 10^{-33} \dots \text{pound (energy)}$$
$$1.782\ 675\ 9 \times 10^{-33} \dots \text{kilogram (energy)}$$

One MEGAELECTRONVOLT is equal to:

$$1\ 000\ 000 \dots \text{electronvolts*}$$
$$1\ 000 \dots \text{kiloelectronvolts*}$$
$$0.001\ 073\ 535\ 4 \dots \text{unified atomic mass}$$
$$\text{unit (energy)}$$

$1.602\ 189\ 2 \times 10^{-13}$ . . . . joule
$1.181\ 714\ 1 \times 10^{-13}$ . . . . foot pound-force

$5.968\ 253 \times 10^{-20}$ . . . . horsepower-hour
$1.782\ 675\ 9 \times 10^{-27}$ . . . . gram (energy)
$3.930\ 128 \times 10^{-30}$ . . . . pound (energy)
$1.782\ 675\ 9 \times 10^{-30}$ . . . . kilogram (energy)

One UNIFIED ATOMIC MASS UNIT (ENERGY) is equal to:

$9.315\ 016 \times 10^{8}$ . . . . electronvolts
$931\ 501.6$ . . . . kiloelectronvolts
$931.501\ 6$ . . . . megaelectronvolts
$1.492\ 441\ 8 \times 10^{-10}$ . . . . joule
$1.100\ 768\ 6 \times 10^{-10}$ . . . . foot pound-force

$5.559\ 437 \times 10^{-17}$ . . . . horsepower-hour
$1.660\ 565\ 5 \times 10^{-24}$ . . . . gram (energy)
$3.660\ 920 \times 10^{-27}$ . . . . pound (energy)
$1.660\ 565\ 5 \times 10^{-27}$ . . . . kilogram (energy)

One JOULE is equal to:

$6.241\ 460 \times 10^{18}$ . . . . electronvolts
$6.241\ 460 \times 10^{15}$ . . . . kiloelectronvolts
$6.241\ 460 \times 10^{12}$ . . . . megaelectronvolts
$6.700\ 429 \times 10^{9}$ . . . . unified atomic mass
units (energy)
$0.737\ 562\ 149\ 3$ . . . . foot pound-force

$3.725\ 061\ 360 \times 10^{-7}$ . . . . . . horsepower-hour
$1.112\ 650\ 06 \times 10^{-14}$ . . . . gram (energy)
$2.452\ 973\ 48 \times 10^{-17}$ . . . . pound (energy)
$1.112\ 650\ 06 \times 10^{-17}$ . . . . . . kilogram (energy)

One FOOT POUND-FORCE is equal to:

$8.462\ 284 \times 10^{18}$ . . . . electronvolts
$8.462\ 284 \times 10^{15}$ . . . . kiloelectronvolts
$8.462\ 284 \times 10^{12}$ . . . . megaelectronvolts
$9.084\ 561 \times 10^{9}$ . . . . unified atomic mass
units (energy)

1.355 817 948 331 400 4 . . . . .joules*

5.050 505 051 × 10^{-7}. . . . . . . horsepower-hour
1.508 550 92 × 10^{-14} . . . . gram (energy)
3.325 785 48 × 10^{-17} . . . . pound (energy)
1.508 550 92 × 10^{-17} . . . . . . .kilogram (energy)

One HORSEPOWER-HOUR is equal to:

1.675 532 2 × 10^{25} . . . . electronvolts
1.675 532 2 × 10^{22} . . . . kiloelectronvolts
1.675 532 2 × 10^{19} . . . . . . .megaelectronvolts
1.798 743 1 × 10^{16} . . . . . . unified atomic mass
units (energy)
2 684 519.537 696 172 792 . . . . .joules*

1 980 000. . . . . .foot pounds-force*
2.986 930 81 × 10^{-8}. . . . . gram (energy)
6.585 055 2 × 10^{-11} . . . . pound (energy)
2.986 930 81 × 10^{-11} . . . . . . .kilogram (energy)

One GRAM (ENERGY) is equal to:

5.609 545 × 10^{32} . . . . electronvolts
5.609 545 × 10^{29} . . . . kiloelectronvolts
5.609 545 × 10^{26} . . . . .megaelectronvolts
6.022 045 × 10^{23} . . . . unified atomic mass
units (energy)
8.987 551 79 × 10^{13}. . . . . .joules

6.628 878 01 × 10^{13}. . . . . . .foot pounds-force
33 479 181.9 . . . . .horsepower-hours
0.002 204 622 622 . . . . pound (energy)
0.001. . . . . kilogram (energy)*

One POUND (ENERGY) is equal to:

2.544 447 × 10^{35} . . . . electronvolts
2.544 447 × 10^{32} . . . . kiloelectronvolts
2.544 447 × 10^{29} . . . . .megaelectronvolts
2.731 554 × 10^{26} . . . . unified atomic mass
units (energy)

$4.076\ 684\ 92 \times 10^{16}$. . . . . .joules

$3.006\ 808\ 49 \times 10^{16}$. . . . . . .foot pounds-force
$1.518\ 590\ 15 \times 10^{10}$. . . . . . .horsepower-hours
$453.592\ 37$ . . . . .grams (energy)*
$0.453\ 592\ 37$. . . . . .kilogram (energy)*

One KILOGRAM (ENERGY) is equal to:

$5.609\ 545 \times 10^{35}$ . . . . electronvolts
$5.609\ 545 \times 10^{32}$ . . . . kiloelectronvolts
$5.609\ 545 \times 10^{29}$ . . . . .megaelectronvolts
$6.022\ 045 \times 10^{26}$ . . . . unified atomic mass
units (energy)
$8.987\ 551\ 79 \times 10^{16}$. . . . . .joules

$6.628\ 878\ 01 \times 10^{16}$ . . . . . . .foot pounds-force
$3.347\ 918\ 19 \times 10^{10}$. . . . . . .horsepower-hours
$1\ 000$. . . . . .grams (energy)*
$2.204\ 622\ 622$. . . . . .pounds (energy)

ADDITIONAL ATOMIC ENERGY UNITS

1 million-
electronvolt = $1\ 000\ 000$ . . . . electronvolts*
              = 1 . . . . megaelectronvolt*
(in tables)
1 giga-
electronvolt = $1\ 000\ 000\ 000$ . . . . electronvolts*
1 billion-
electronvolt = $1\ 000\ 000\ 000$ . . . . electronvolts*

ADDITIONAL ATOMIC MASS AND ENERGY UNITS

electron rest
mass   = $9.109\ 534 \times 10^{-28}$ . . . . gram
          = $0.000\ 548\ 580\ 26$. . . .unified atomic
mass unit
          = $0.511\ 003\ 4$ . . . . . . megaelectronvolt
          = $8.187\ 241 \times 10^{-14}$ . . . . joule

proton rest
  mass   = $1.672\ 648\ 5 \times 10^{-24}$ . . . . gram
              = $1.007\ 276\ 470$ . . . . . . unified atomic
                                               mass units
              = $938.279\ 6$ . . . . .megaelectronvolts
              = $1.503\ 301\ 5 \times 10^{-10}$ . . . . joule

neutron rest
  mass   = $1.674\ 954\ 3 \times 10^{-24}$ . . . . gram
              = $1.008\ 665\ 012$ . . . . . . unified atomic
                                               mass units
              = $939.573\ 1$ . . . . .megaelectronvolts
              = $1.505\ 373\ 9 \times 10^{-10}$ . . . . joule

Density and Concentration

One GRAM PER CUBIC METER or KILOLITER is equal to:

0.436 995 724 0 . . . . gran per cubic foot

0.058 417 831 16 . . . . grain per gallon

0.001 685 554 936 . . . . . . pound per cubic yard

0.001 . . . . .kilogram per cubic meter
or gram per liter*

0.000 998 847 369 2 . . . . . .ounce per cubic foot

0.000 252 891 044 0 . . . . . grain per cubic inch

0.000 133 526 471 2 . . . . .ounce per gallon

0.000 062 427 960 58 . . . . pound per cubic foot

0.000 008 345 404 452 . . . . . . pound per gallon

0.000 001 . . . . .gram per cubic centimeter or milliliter*

$8.427\ 774\ 678 \times 10^{-7}$ . . . . .short ton per
cubic yard

$7.524\ 798\ 819 \times 10^{-7}$ . . . . long ton per
cubic yard

$5.780\ 366\ 720 \times 10^{-7}$ . . . . .ounce per cubic inch

$3.612\ 729\ 200 \times 10^{-8}$ . . . . pound per cubic inch

One GRAIN PER CUBIC FOOT is equal to:

2.288 351 911 . . . . .grams per cubic meter
or kiloliter

0.133 680 555 6 . . . . grain per gallon

0.003 857 142 857 . . . . . . pound per cubic yard

0.002 288 351 911 . . . . kilogram/cubic meter or
gram/liter

0.002 285 714 286 . . . . . . .ounce per cubic foot

0.000 578 703 703 7 . . . . . grain per cubic inch

0.000 305 555 555 6 . . . . .ounce per gallon

0.000 142 857 142 9. . . . . pound per cubic foot
0.000 019 097 222 22. . . . . . . pound per gallon
0.000 002 288 351 911 . . . . .gram/cu centimeter
or milliliter

0.000 001 928 571 429 . . . . .short ton per
cubic yard
0.000 001 721 938 776 . . . . long ton per
cubic yard
0.000 001 322 751 323 . . . .ounce per cubic inch
8.267 195 767 × 10^{-8}. . . . pound per cubic inch

One GRAIN PER GALLON is equal to:

17.118 061 05. . . . . .grams per cubic meter
or kiloliter
7.480 519 481. . . . . .grains per cubic foot
0.028 853 432 28 . . . . . . pound per cubic yard
0.017 118 061 05 . . . . . kilogram/cubic meter or
gram/liter
0.017 098 330 24 . . . . .ounce per cubic foot

0.004 329 004 329 . . . . . . grain per cubic inch
0.002 285 714 286 . . . . .ounce per gallon
0.001 068 645 640 . . . . . pound per cubic foot
0.000 142 857 142 9. . . . . . . pound per gallon
0.000 017 118 061 05. . . .gram/cu centimeter or
milliliter

0.000 014 426 716 14. . . . . .short ton per
cubic yard
0.000 012 880 996 55. . . . . long ton per
cubic yard
0.000 009 894 867 038 . . . .ounce per cubic inch
6.184 291 899 × 10^{-7}. . . . pound per cubic inch

One POUND PER CUBIC YARD is equal to:

593.276 421 3. . . . . .grams per cubic meter
or kiloliter
259.259 259 3 . . . . .grains per cubic foot

34.657 921 81 . . . . .grains per gallon
0.593 276 421 3 . . . . . . kilogram/cubic meter or
gram/liter
0.592 592 592 6 . . . . .ounce per cubic foot

0.150 034 293 6 . . . . grain per cubic inch
0.079 218 107 00 . . . . .ounce per gallon
0.037 037 037 04 . . . . pound per cubic foot
0.004 951 131 687 . . . . pound per gallon
0.000 593 276 421 3 . . . . .gram/cu centimeter or
milliliter

0.000 5 . . . . .short ton per cubic yard*
0.000 446 428 571 4 . . . . long ton per cubic yard
0.000 342 935 528 1 . . . . . .ounce per cubic inch
0.000 021 433 470 51 . . . . pound per cubic inch

One **KILOGRAM PER CUBIC METER** or **GRAM PER LITER** is equal to:

1 000 . . . . .grams per cubic meter
or kiloliter*
436.995 724 0 . . . . .grains per cubic foot
58.417 831 16 . . . . .grains per gallon
1.685 554 936 . . . . .pounds per cubic yard
0.998 847 369 2 . . . . .ounce per cubic foot

0.252 891 044 0 . . . . grain per cubic inch
0.133 526 471 2 . . . . .ounce per gallon
0.062 427 960 58 . . . . pound per cubic foot
0.008 345 404 452 . . . . pound per gallon
0.001 . . . . gram per cubic centimeter or
milliliter*

0.000 842 777 467 8 . . . . .short ton per
cubic yard
0.000 752 479 881 9 . . . . long ton per
cubic yard
0.000 578 036 672 0 . . . . . .ounce per cubic inch
0.000 036 127 292 00 . . . . pound per cubic inch

One **OUNCE PER CUBIC FOOT** is equal to:

1 001.153 961 . . . . grams per cubic meter or
kiloliter

437.5 . . . . grains per cubic foot*

58.485 243 06. . . . . grains per gallon

1.687 5 . . . . pounds per cubic yard*

1.001 153 961 . . . . kilograms/cubic meter or
grams/liter

0.253 182 870 4 . . . . grain per cubic inch

0.133 680 555 6 . . . . ounce per gallon

0.062 5 . . . . pound per cubic foot*

0.008 355 034 722 . . . . pound per gallon

0.001 001 153 961 . . . . . . gram/cu centimeter or
milliliter

0.000 843 75. . . . . short ton per
cubic yard*

0.000 753 348 214 3. . . . . long ton per
cubic yard

0.000 578 703 703 7. . . . . . ounce per cubic inch

0.000 036 168 981 48. . . . pound per cubic inch

One **GRAIN PER CUBIC INCH** is equal to:

3 954.272 101 . . . . grams per cubic meter
or kiloliter

1 728. . . . . grains per cubic foot*

231. . . . . grains per gallon*

6.665 142 857. . . . . pounds per cubic yard

3.954 272 101 . . . . kilograms/cubic meter or
grams/liter

3.949 714 286. . . . . ounces per cubic foot

0.528. . . . . ounce per gallon*

0.246 857 142 9 . . . . pound per cubic foot

0.033. . . . . pound per gallon*

0.003 954 272 101 . . . . . . gram/cu centimeter or
milliliter

0.003 332 571 429 . . . . short ton per
cubic yard
0.002 975 510 204 . . . . long ton per
cubic yard
0.002 285 714 286 . . . . ounce per cubic inch
0.000 142 857 142 9. . . . . pound per cubic inch

One OUNCE PER GALLON is equal to:

7 489.151 707 . . . . grams per cubic meter or
kiloliter
3 272.727 273 . . . . grains per cubic foot
437.5 . . . . grains per gallon*
12.623 376 62. . . . . .pounds per cubic yard
7.489 151 707 . . . . kilograms/cubic meter or
grams/liter

7.480 519 481 . . . . ounces per cubic foot
1.893 939 394. . . . . .grains per cubic inch
0.467 532 467 5 . . . . pound per cubic foot
0.062 5 . . . . pound per gallon*
0.007 489 151 707 . . . . . .gram/cu centimeter or
milliliter

0.006 311 688 312 . . . . short ton per
cubic yard
0.005 635 435 993 . . . . long ton per
cubic yard
0.004 329 004 329 . . . . ounce per cubic inch
0.000 270 562 770 6. . . . . pound per cubic inch

One POUND PER CUBIC FOOT is equal to:

16 018.463 37 . . . . grams per cubic meter or
kiloliter
7 000. . . . . grains per cubic foot*
935.763 888 9. . . . . .grains per gallon
27. . . . . pounds per cubic yard*
16.018 463 37. . . . . kilograms/cubic meter or
grams/liter

16. . . . . ounces per cubic foot*
4.050 925 926. . . . . .grains per cubic inch
2.138 888 889. . . . . .ounces per gallon
0.133 680 555 6 . . . . pound per gallon
0.016 018 463 37 . . . . . . .gram/cu centimeter or
milliliter

0.013 5. . . . . .short ton per
cubic yard*
0.012 053 571 43 . . . . long ton per
cubic yard
0.009 259 259 259 . . . . . . .ounce per cubic inch
0.000 578 703 703 7. . . . . pound per cubic inch

One **POUND PER GALLON** is equal to:

119 826.427 3 . . . . grams per cubic meter or
kiloliter
52 363.636 36 . . . . .grains per cubic foot
7 000. . . . . .grains per gallon*
201.974 026 0. . . . . .pounds per cubic yard
119.826 427 3. . . . . .kilograms/cubic meter or
grams/liter

119.688 311 7. . . . . ounces per cubic foot
30.303 030 30. . . . . .grains per cubic inch
16. ounces per gallon*
7.480 519 481. . . . . .pounds per cubic foot
0.119 826 427 3 . . . . .gram/cu centimeter
or milliliter

0.100 987 013 0 . . . . .short ton per
cubic yard
0.090 166 975 88 . . . . long ton per
cubic yard
0.069 264 069 26 . . . . .ounce per cubic inch
0.004 329 004 329 . . . . . . pound per cubic inch

One **GRAM PER CUBIC CENTIMETER** or **MILLILITER** is equal to:

1 000 000. . . . . .grams per cubic meter or
kiloliter*

436 995.724 0 . . . . grains per cubic foot
58 417.831 16 . . . . grains per gallon
1 685.554 936 . . . . pounds per cubic yard
1 000 . . . . kilograms per cubic meter or grams
per liter*

998.847 369 2 . . . . ounces per cubic foot
252.891 044 0 . . . . grains per cubic inch
133.526 471 2 . . . . ounces per gallon
62.427 960 58 . . . . pounds per cubic foot
8.345 404 452 . . . . pounds per gallon

0.842 777 467 8 . . . . short ton per
cubic yard
0.752 479 881 9 . . . . long ton per
cubic yard
0.578 036 672 0 . . . . ounce per cubic inch
0.036 127 292 00 . . . . pound per cubic inch

One SHORT TON PER CUBIC YARD is equal to:

1 186 552.843 . . . . grams per cubic meter or
kiloliter
518 518.518 5 . . . . grains per cubic foot
69 315.843 62 . . . . grains per gallon
2 000 . . . . pounds per cubic yard*
1 186.552 843 . . . . kilograms/cubic meter or
grams/liter

1 185.185 185 . . . . ounces per cubic foot
300.068 587 1 . . . . grains per cubic inch
158.436 214 0 . . . . ounces per gallon
74.074 074 07 . . . . pounds per cubic foot
9.902 263 374 . . . . pounds per gallon

1.186 552 843 . . . . grams/cubic centimeter or
milliliter
0.892 857 142 9 . . . . long ton per cubic yard
0.685 871 056 2 . . . . ounce per cubic inch
0.042 866 941 02 . . . . pound per cubic inch

One LONG TON PER CUBIC YARD is equal to:

1 328 939.184 . . . . grams per cubic meter or
kiloliter
580 740.740 7 . . . . grains per cubic foot
77 633.744 86 . . . . grains per gallon
2 240 . . . . pounds per cubic yard*
1 328.939 184 . . . . kilograms/cubic meter or
grams/liter

1 327.407 407 . . . . ounces per cubic foot
336.076 817 6 . . . . grains per cubic inch
177.448 559 7 . . . . ounces per gallon
82.962 962 96 . . . . pounds per cubic foot
11.090 534 98 . . . . .pounds per gallon

1.328 939 184 . . . . grams/cubic centimeter or
milliliter
1.12 . . . . short tons per cubic yard*
0.768 175 583 0 . . . . .ounce per cubic inch
0.048 010 973 94 . . . . pound per cubic inch

One OUNCE PER CUBIC INCH is equal to:

1 729 994.044 . . . . grams per cubic meter
or kiloliter
756 000 . . . . grains per cubic foot*
101 062.5 . . . . grains per gallon*
2 916 . . . . .pounds per cubic yard*
1 729.994 044 . . . . kilograms/cubic meter or
grams/liter

1 728 . . . . ounces per cubic foot*
437.5 . . . . grains per cubic inch*
231 . . . . ounces per gallon*
108 . . . . pounds per cubic foot*
14.437 5 . . . . .pounds per gallon*

1.729 994 044 . . . . grams/cubic centimeter
or milliliter
1.458 . . . . short tons per cubic yard*
1.301 785 714 . . . . long tons per cubic yard
0.062 5 . . . . .pound per cubic inch*

One POUND PER CUBIC INCH is equal to:

27 679 904.71. . . . . grams per cubic meter
or kiloliter
12 096 000. . . . . grains per cubic foot*
1 617 000. . . . . grains per gallon*
46 656. . . . . pounds per cubic yard*
27 679.904 71 . . . . .kilograms/cubic meter
or grams/liter

27 648. . . . . ounces per cubic foot*
7 000. . . . . grains per cubic inch*
3 696. . . . . ounces per gallon*
1 728. . . . . pounds per cubic foot*
231. . . . . .pounds per gallon*

27.679 904 71. . . . . grams/cubic centimeter
or milliliter
23.328. . . . . short tons per cubic yard*
20.828 571 43. . . . . long tons per cubic yard
16. ounces per cubic inch*

ADDITIONAL UNITS OF DENSITY AND CONCENTRATION

Metric

1 microgram/cubic meter or kiloliter	=	0.000 000 001. . . . kilogram/cu meter*
1 microgram/cubic decimeter or liter	=	0.000 001 . . . . . kilogram/cubic meter*
1 milligram/cubic meter or kiloliter	=	0.000 001 . . . . . kilogram/cubic meter*
1 microgram per cubic centimeter or milliliter	=	1. . . . . . . gram/cubic meter or kiloliter* (in tables)
	=	0.001. . . . . . . kilogram per cubic meter*
1 milligram/cubic decimeter or liter	=	1. . . . . . . gram/cubic meter or kiloliter* (in tables)
	=	0.001. . . . . . . kilogram per cubic meter*
1 milligram per cubic centimeter or milliliter	=	1. . . . . kilogram per cubic meter* (in tables)

1 gram per cubic decimeter	=	1. . . . . kilogram per cubic meter* (in tables)
1 kilogram per kiloliter	=	1. . . . . kilogram per cubic meter* (in tables)
1 kilogram/cubic decimeter or liter	=	1. . . . . gram per cubic centimeter or milliliter* (in tables)
	=	1 000 . . . . kilograms/cubic meter*

Customary

1 ounce per petroleum barrel	=	0.178 313 1 . . . . kilogram/cubic meter
1 pound per petroleum barrel	=	2.853 010 . . . . . .kilograms/cubic meter
1 slug per cubic foot	=	515.378 8 . . . . .kilograms/cubic meter

Electricity and Magnetism

ELECTRIC CURRENT

One **STATAMPERE** is equal to:

$$3.335\ 640\ 95 \times 10^{-10} \dots \dots \dots \dots \text{ampere}$$
$$3.335\ 640\ 95 \times 10^{-11} \dots \dots \dots \dots \text{abampere}$$

One **AMPERE** is equal to:

$$2.997\ 924\ 58 \times 10^{9} \dots \dots \dots \dots \text{statamperes}$$
$$0.1 \dots \dots \dots \dots \dots \dots \dots \dots \text{abampere*}$$

One **ABAMPERE** is equal to:

$$2.997\ 924\ 58 \times 10^{10} \dots \dots \dots \dots \text{statamperes}$$
$$10 \dots \dots \dots \dots \dots \dots \dots \dots \text{amperes*}$$

ELECTRIC CHARGE, QUANTITY OF ELECTRICITY

One **STATCOULOMB** is equal to:

$$3.335\ 640\ 95 \times 10^{-10} \dots \dots \dots \dots \text{coulomb}$$
$$3.335\ 640\ 95 \times 10^{-11} \dots \dots \dots \dots \text{abcoulomb}$$

One **COULOMB** is equal to:

$$2.997\ 924\ 58 \times 10^{9} \dots \dots \dots \dots \text{statcoulombs}$$
$$0.1 \dots \dots \dots \dots \dots \dots \dots \dots \text{abcoulomb*}$$

One **ABCOULOMB** is equal to:

$$2.997\ 924\ 58 \times 10^{10} \dots \dots \dots \dots \text{statcoulombs}$$
$$10 \dots \dots \dots \dots \dots \dots \dots \dots \text{coulombs*}$$

ELECTRIC POTENTIAL, POTENTIAL DIFFERENCE, ELECTROMOTIVE FORCE

One ABVOLT is equal to:

$$0.000\ 000\ 01 \dots\dots\dots\dots\dots\dots \text{volt*}$$
$$3.335\ 640\ 95 \times 10^{-11} \dots\dots\dots\dots \text{statvolt}$$

One VOLT is equal to:

$$100\ 000\ 000 \dots\dots\dots\dots\dots\dots\dots \text{abvolts*}$$
$$0.003\ 335\ 640\ 95 \dots\dots\dots\dots \text{statvolt}$$

One STATVOLT is equal to:

$$2.997\ 924\ 58 \times 10^{10} \dots\dots\dots\dots \text{abvolts}$$
$$299.792\ 458 \dots\dots\dots\dots\dots\dots \text{volts}$$

ELECTRIC RESISTANCE

One ABOHM is equal to:

$$0.000\ 000\ 001 \dots\dots\dots\dots\dots \text{ohm*}$$
$$1.112\ 650\ 06 \times 10^{-21} \dots\dots\dots \text{statohm}$$

One OHM is equal to:

$$1\ 000\ 000\ 000 \dots\dots\dots\dots\dots\dots \text{abohms*}$$
$$1.112\ 650\ 06 \times 10^{-12} \dots\dots\dots \text{statohm}$$

One STATOHM is equal to:

$$8.987\ 551\ 79 \times 10^{20} \dots\dots\dots\dots \text{abohms}$$
$$8.987\ 551\ 79 \times 10^{11} \dots\dots\dots\dots \text{ohms}$$

ELECTRIC CAPACITANCE

One STATFARAD is equal to:

$$1.112\ 650\ 06 \times 10^{-12} \dots\dots\dots\dots \text{farad}$$
$$1.112\ 650\ 06 \times 10^{-21} \dots\dots\dots\dots \text{abfarad}$$

One FARAD is equal to:

$$8.987\ 551\ 79 \times 10^{11} \ldots \ldots \ldots \ldots \text{statfarads}$$
$$0.000\ 000\ 001 \ldots \ldots \ldots \ldots \ldots \text{.abfarad*}$$

One ABFARAD is equal to:

$$8.987\ 551\ 79 \times 10^{20} \ldots \ldots \ldots \ldots \text{statfarads}$$
$$1\ 000\ 000\ 000 \ldots \ldots \ldots \ldots \ldots \ldots \ldots \ldots \text{.farads*}$$

ELECTRIC INDUCTANCE

One ABHENRY is equal to:

$$0.000\ 000\ 001 \ldots \ldots \ldots \ldots \ldots \text{.henry*}$$
$$1.112\ 650\ 06 \times 10^{-21} \ldots \ldots \ldots \ldots \text{stathenry}$$

One HENRY is equal to:

$$1\ 000\ 000\ 000 \ldots \ldots \ldots \ldots \ldots \ldots \ldots \text{abhenries*}$$
$$1.112\ 650\ 06 \times 10^{-12} \ldots \ldots \ldots \ldots \text{stathenry}$$

One STATHENRY is equal to:

$$8.987\ 551\ 79 \times 10^{20} \ldots \ldots \ldots \ldots \text{abhenries}$$
$$8.987\ 551\ 79 \times 10^{11} \ldots \ldots \ldots \ldots \text{.henries}$$

ELECTRIC CONDUCTANCE

One STATSIEMENS is equal to:

$$1.112\ 650\ 06 \times 10^{-12} \ldots \ldots \ldots \ldots \text{siemens}$$
$$1.112\ 650\ 06 \times 10^{-21} \ldots \ldots \ldots \ldots \text{absiemens}$$

One SIEMENS is equal to:

$$8.987\ 551\ 79 \times 10^{11} \ldots \ldots \ldots \ldots \text{statsiemens}$$
$$0.000\ 000\ 001 \ldots \ldots \ldots \ldots \ldots \text{absiemens*}$$

One ABSIEMENS is equal to:

$$8.987\ 551\ 79 \times 10^{20} \ldots \ldots \ldots \ldots \text{statsiemens}$$
$$1\ 000\ 000\ 000 \ldots \ldots \ldots \ldots \ldots \ldots \ldots \text{.siemens*}$$

MAGNETIC FLUX

One MAXWELL is equal to:

0.000 000 01. .weber*

One WEBER is equal to:

100 000 000. . . . . .maxwells*

MAGNETIC FLUX DENSITY, MAGNETIC INDUCTION

One GAUSS is equal to:

0.000 1 . . . . .tesla*

One TESLA is equal to:

10 000. . . . . gausses*

MAGNETIC FIELD STRENGTH

One AMPERE PER METER corresponds to:

0.012 566 370 61 . . . . .oersted

One OERSTED corresponds to:

79.577 471 55. . . . . amperes per meter

MAGNETOMOTIVE FORCE, MAGNETIC POTENTIAL DIFFERENCE

One GILBERT corresponds to:

0.795 774 715 5 . . . . .ampere

One AMPERE corresponds to:

1.256 637 061. . . . . .gilberts

ADDITIONAL UNITS OF ELECTRICITY AND MAGNETISM

(Units with the prefix "ab" are electromagnetic units abbreviated EMU, and units with the prefix "stat" are elestrostatic units abbreviated ESU.)

ELECTRIC CURRENT

1 electromagnetic unit, EMU	=	1. . . . . abampere* (in	tables)
1 electrostatic unit, ESU	=	1. . . . . statampere* (in	tables)
1 microampere	=	0.000 001 . . . . ampere*	
1 milliampere	=	0.001. . . . . ampere*	

ELECTRIC CHARGE, QUANTITY OF ELECTRICITY

1 electromagnetic unit, EMU	=	1. . . . . abcoulomb* (in	tables)
1 electrostatic unit, ESU	=	1. . . . . statcoulomb* (in	tables)
1 ampere-second	=	1. . . . . .coulomb* (in tables)	
1 ampere-hour	=	3 600. . . . . coulombs*	

ELECTRIC POTENTIAL, POTENTIAL DIFFERENCE, ELEC-TROMOTIVE FORCE

1 electromagnetic unit, EMU	=	1. . . . . abvolt* (in	tables)
1 electrostatic unit, ESU	=	1. . . . . statvolt* (in	tables)
1 microvolt	=	0.000 001 . . . . volt*	
1 millivolt	=	0.001. . . . . volt*	
1 watt per ampere	=	1. . . . . .volt* (in	tables)
1 kilovolt	=	1 000. .volts*	

ELECTRIC RESISTANCE

1 electromagnetic unit, EMU	=	1 . . . . abohm* (in tables)
1 electrostatic unit, ESU	=	1 . . . . statohm* (in tables)
1 microhm	=	0.000 001 . . . . ohm*
1 volt per ampere	=	1 . . . . ohm* (in tables)
1 megohm	= 1 000 000	. . . . ohms*

ELECTRIC CAPACITANCE

1 electromagnetic unit, EMU	=	1. . . . . abfarad* (in tables)
1 electrostatic unit, ESU	=	1. . . . . .statfarad* (in tables)
1 picofarad	=	0.000 000 000 001 . . . . farad*
1 microfarad	=	0.000 001 . . . . farad*
1 coulomb per volt	=	1. farad* (in tables)

ELECTRIC INDUCTANCE

1 electromagnetic unit, EMU	=	1. . . . . abhenry* (in tables)
1 electrostatic unit, ESU	=	1. . . . . stathenry* (in tables)
1 microhenry	=	0.000 001 . . . . henry*
1 millihenry	=	0.001. . . . . henry*
1 volt-second per ampere	=	1. . . . . .henry* (in tables)

ELECTRIC CONDUCTANCE

1 electromagnetic unit, EMU	=	1. . . . . .absiemens* (in tables)

1 electrostatic
  unit, ESU   =   1................statsiemens* (in tables)
1 statmho   =   1................statsiemens* (in tables)
1 mho   =   1................ siemens* (in tables)
1 ampere per volt =   1................ siemens* (in tables)
1 abmho   =   1................ absiemens* (in tables)

MAGNETIC FLUX

1 electromagnetic
  unit, EMU   =   1.................maxwell* (in tables)
1 line   =   1.................maxwell* (in tables)
1 unit pole   = $1.256\ 637\ 061 \times 10^{-7}$.......weber
1 kiloline   =   1 000...................maxwells*
1 megaline   = 1 000 000...................maxwells*
1 volt-second   =   1................... weber* (in tables)

MAGNETIC FLUX DENSITY, MAGNETIC INDUCTION

1 electromagnetic
  unit, EMU   =   1..................... gauss* (in tables)
1 gamma   =   0.000 000 001.............tesla*
1 maxwell per
  square centi-
  meter   =   1............... gauss* (in tables)
1 line per
  square centi-
  meter   =   1............... gauss* (in tables)
1 weber per
  square meter   =   1...............tesla* (in tables)
1 weber per
  square centi-
  meter   =   10 000..................... teslas*

MAGNETIC FIELD STRENGTH

1 electromagnetic
  unit, EMU   =   1..............oersted* (in tables)

MAGNETOMOTIVE FORCE, MAGNETIC POTENTIAL DIF-FERENCE

1 electromagnetic unit, EMU	=	1. . . . . gilbert* (in tables)
1 oersted- centimeter	=	1. . . . . gilbert* (in tables)

ADDITIONAL CATEGORIES AND UNITS OF ELECTRICITY AND MAGNETISM

ELECTRIC LINEAR CURRENT DENSITY Basic unit: ampere per meter

ELECTRIC CURRENT DENSITY Basic unit: ampere per square meter

1 statampere per square centi- meter	=	0.000 003 335 640 95. . . . ampere per sq meter
1 abampere per square centi- meter	=	100 000. . . . . amperes/square meter*

ELECTRIC FLUX, FLUX OF DISPLACEMENT Basic unit: coulomb

ELECTRIC DIPOLE MOMENT Basic unit: coulomb-meter

ELECTRIC FLUX DENSITY, SURFACE DENSITY OF CHARGE Basic unit: coulomb per square meter

ELECTRIC DISPLACEMENT Basic unit: coulomb per square meter

ELECTRIC POLARIZATION Basic unit: coulomb per square meter

ELECTRIC CHARGE DENSITY, VOLUME DENSITY OF CHARGE Basic unit: coulomb per cubic meter

ELECTRIC FIELD STRENGTH Basic unit: volt per meter

| 1 abvolt per centi-
meter | = | 0.000 001 . . . . volt per meter* |
| 1 statvolt per
centimeter | = | 29 979.245 8 . . . . .volts per meter |

ELECTRIC IMPEDANCE, MODULUS OF IMPEDANCE
Basic unit: ohm

ELECTRIC REACTANCE Basic unit: ohm

ELECTRIC RESISTIVITY Basic unit: ohm-meter

1 abohm-centi- meter	=	0.000 000 000 01 . . . . . . ohm-meter*
1 microhm-centi- meter	=	0.000 000 01 . . . . ohm-meter*
1 ohm-centimeter	=	0.01. . . . . ohm-meter*
1 statohm-centi- meter	=	8.987 551 79 $\times$ 10^9 . . . . .ohm-meters

PERMITTIVITY Basic unit: farad per meter

PERMEABILITY Basic unit: henry per meter

PERMEANCE Basic units: henry, weber per ampere

RELUCTANCE Basic units: reciprocal of henry, ampere per weber

ELECTRIC ADMITTANCE, MODULUS OF ADMITTANCE
Basic unit: siemens

ELECTRIC SUSCEPTANCE Basic unit: siemens

ELECTRIC CONDUCTIVITY Basic unit: siemens per meter

MAGNETIC DIPOLE MOMENT Basic unit: weber-meter

MAGNETIC VECTOR POTENTIAL Basic unit: weber per meter

MAGNETIC POLARIZATION Basic unit: tesla

MAGNETIZATION Basic unit: ampere per meter

MAGNETIC MOMENT, ELECTROMAGNETIC MOMENT
Basic unit: ampere-sq meter

Energy

One ERG is equal to:

$$0.1 \ldots \ldots \ldots \ldots \ldots \ldots \ldots \text{microjoule*}$$
$$0.001 \ 019 \ 716 \ 213 \ \ldots \ldots \text{gram-force centimeter}$$
$$0.000 \ 002 \ 373 \ 036 \ 040 \ldots \ldots \ldots \text{foot poundal}$$
$$0.000 \ 000 \ 1 \ldots \ldots \ldots \ldots \ldots \ldots \text{joule*}$$
$$7.375 \ 621 \ 493 \times 10^{-8} \ldots \ldots \text{foot pound-force}$$

$$2.390 \ 057 \ 361 \times 10^{-8} \ldots \ldots \ldots \ldots \text{calorie}$$
$$1.019 \ 716 \ 213 \times 10^{-8} \ldots \ldots \text{kilogram-force meter}$$
$$0.000 \ 000 \ 000 \ 1 \ldots \ldots \ldots \ldots \ldots \text{kilojoule*}$$
$$9.484 \ 514 \ 148 \times 10^{-11} \ldots \ldots \ldots \ldots \text{Btu}$$
$$2.777 \ 777 \ 778 \times 10^{-11} \ldots \ldots \ldots \text{watt-hour}$$

$$2.390 \ 057 \ 361 \times 10^{-11} \ldots \ldots \ldots \text{kilocalorie}$$
$$0.000 \ 000 \ 000 \ 000 \ 1 \ldots \ldots \ldots \text{megajoule*}$$
$$3.776 \ 726 \ 715 \times 10^{-14} \ldots \ldots \text{metric horsepower-}$$
$$\text{hour}$$
$$3.725 \ 061 \ 360 \times 10^{-14} \ldots \ldots \text{horsepower-hour}$$
$$3.723 \ 562 \ 705 \times 10^{-14} \ldots \ldots \text{electric horsepower-}$$
$$\text{hour}$$
$$2.777 \ 777 \ 778 \times 10^{-14} \ldots \ldots \ldots \text{kilowatt-hour}$$

One MICROJOULE is equal to:

$$10 \ldots \ldots \ldots \ldots \ldots \ldots \ldots \ldots \text{ergs*}$$
$$0.010 \ 197 \ 162 \ 13 \ \ldots \ldots \text{gram-force centimeter}$$
$$0.000 \ 023 \ 730 \ 360 \ 40 \ldots \ldots \ldots \text{foot poundal}$$
$$0.000 \ 001 \ldots \ldots \ldots \ldots \ldots \ldots \text{joule*}$$
$$7.375 \ 621 \ 493 \times 10^{-7} \ldots \ldots \text{foot pound-force}$$

$$2.390 \ 057 \ 361 \times 10^{-7} \ldots \ldots \ldots \ldots \text{calorie}$$
$$1.019 \ 716 \ 213 \times 10^{-7} \ldots \ldots \text{kilogram-force meter}$$
$$0.000 \ 000 \ 001 \ldots \ldots \ldots \ldots \ldots \text{kilojoule*}$$

$9.484\ 514\ 148 \times 10^{-10}$ . . . . Btu

$2.777\ 777\ 778 \times 10^{-10}$ . . . . watt-hour

$2.390\ 057\ 361 \times 10^{-10}$ . . . . .kilocalorie

$0.000\ 000\ 000\ 001$ . . . . megajoule*

$3.776\ 726\ 715 \times 10^{-13}$ . . . . .metric horsepower-
hour

$3.725\ 061\ 360 \times 10^{-13}$ . . . . . . horsepower-hour

$3.723\ 562\ 705 \times 10^{-13}$ . . . . electric horsepower-
hour

$2.777\ 777\ 778 \times 10^{-13}$ . . . . kilowatt-hour

One GRAM-FORCE CENTIMETER is equal to:

980.665. . . . . ergs*

$98.066\ 5$ . . . . microjoules*

$0.002\ 327\ 153\ 389$ . . . . .foot poundal

$0.000\ 098\ 066\ 5$ . . . . joule*

$0.000\ 072\ 330\ 138\ 51$ . . . . foot pound-force

$0.000\ 023\ 438\ 456\ 02$ . . . . calorie

$0.000\ 01$ . . . . .kilogram-force meter*

$0.000\ 000\ 098\ 066\ 5$. . . . . .kilojoule*

$9.301\ 131\ 066 \times 10^{-8}$ . . . . Btu

$2.724\ 069\ 444 \times 10^{-8}$ . . . . watt-hour

$2.343\ 845\ 602 \times 10^{-8}$ . . . . .kilocalorie

$0.000\ 000\ 000\ 098\ 066\ 5$ . . . . megajoule*

$3.703\ 703\ 704 \times 10^{-11}$ . . . . .metric horsepower-
hour

$3.653\ 037\ 299 \times 10^{-11}$ . . . . . . horsepower-hour

$3.651\ 567\ 620 \times 10^{-11}$ . . . . electric horsepower-
hour

$2.724\ 069\ 444 \times 10^{-11}$ . . . . kilowatt-hour

One FOOT POUNDAL is equal to:

$421\ 401.100\ 938\ 048$ . . . . ergs*

$42\ 140.110\ 093\ 804\ 8$ . . . . microjoules*

$429.709\ 534\ 8$. . . . . gram-force centimeters

0.042 140 110 093 804 8 . . . . joule*

0.031 080 950 17 . . . . foot pound-force

0.010 071 728 03 . . . . calorie

0.004 297 095 348 . . . . kilogram-force
meter

0.000 042 140 110 093 804 8 . . . . . . kilojoule*

0.000 039 967 847 04 . . . . Btu

0.000 011 705 586 137 168 . . . . . . watt-hour*

0.000 010 071 728 03 . . . . kilocalorie

0.000 000 042 140 110 093 804 8 . . . . . megajoule*

1.591 516 796 × 10^{-8} . . . . metric horse-
power- hour

1.569 744 958 × 10^{-8} . . . . horsepower-
hour

1.569 113 423 × 10^{-8} . . . . electric horsepower-hour

0.000 000 011 705 586 137 168 . . . . kilowatt-
hour*

One JOULE is equal to:

10 000 000 . . . . ergs*

1 000 000 . . . . microjoules*

10 197.162 13 . . . . gram-force centimeters

23.730 360 40 . . . . foot poundals

0.737 562 149 3 . . . . foot pound-force

0.239 005 736 1 . . . . calorie

0.101 971 621 3 . . . . kilogram-force
meter

0.001 . . . . kilojoule*

0.000 948 451 414 8 . . . . Btu

0.000 277 777 777 8 . . . . watt-hour

0.000 239 005 736 1 . . . . kilocalorie

0.000 001 . . . . megajoule*

3.776 726 715 × 10^{-7} . . . . . metric horsepower-
hour

3.725 061 360 × 10^{-7} . . . . horsepower-hour
3.723 562 705 × 10^{-7} . . . . . electric horsepower-
hour
2.777 777 778 × 10^{-7} . . . . kilowatt-hour

One FOOT POUND-FORCE is equal to:

13 558 179.483 314 004 . . . . ergs*
1 355 817.948 331 400 4 . . . . microjoules*
13 825.495 437 6 . . . . gram-force centimeters*
32.174 048 56 . . . . foot poundals
1.355 817 948 331 400 4 . . . . .joules*

0.324 048 266 8 . . . . calorie
0.138 254 954 376 . . . . kilogram-force
meter*
0.001 355 817 948 331 400 4 . . . . . . kilojoule*
0.001 285 927 451 . . . . Btu
0.000 376 616 096 8 . . . . watt-hour

0.000 324 048 266 8 . . . . .kilocalorie
0.000 001 355 817 948 331 400 4 . . . . .mega-
joule*
5.120 553 866 × 10^{-7} . . . . .metric horsepower-hour
5.050 505 051 × 10^{-7} . . . . horsepower-
hour
5.048 473 147 × 10^{-7} . . . . electric horsepower-hour

3.766 160 968 × 10^{-7} . . . . kilowatt-hour

One CALORIE is equal to:

41 840 000 . . . . ergs*
4 184 000 . . . . microjoules*
42 664.926 35 . . . . gram-force centimeters
99.287 827 93 . . . . foot poundals
4.184 . . . . .joules*

```
3.085 960 033 . . . . . . . . . . . . foot pounds-force
0.426 649 263 5 . . . . . . . . . . . . .kilogram-force
                                                  meter
0.004 184 . . . . . . . . . . . . . . . . . . . .kilojoule*
0.003 968 320 719 . . . . . . . . . . . . . . . . Btu
0.001 162 222 222 . . . . . . . . . . . . . watt-hour

0.001 . . . . . . . . . . . . . . . . . . . . . kilocalorie*
0.000 004 184 . . . . . . . . . . . . . . . megajoule*
0.000 001 580 182 457 . . . . . . . metric horse-
                                          power-hour
0.000 001 558 565 673 . . . . . . . . horsepower-
                                                  hour
0.000 001 557 938 636 . . . . . . . electric horse-
                                          power-hour

0.000 001 162 222 222 . . . . . . . kilowatt-hour
```

One KILOGRAM-FORCE METER is equal to:

```
98 066 500 . . . . . . . . . . . . . . . . . . . . . . . . . . . . . ergs*
 9 806 650 . . . . . . . . . . . . . . . . . . . . . . . . . microjoules*
   100 000 . . . . . . . . . . . . . . . . . . . . gram-force centi-
                                                           meters*
  232.715 338 9 . . . . . . . . . . . . . . . foot poundals
    9.806 65 . . . . . . . . . . . . . . . . . . . . . . . joules*

    7.233 013 851 . . . . . . . . . . . foot pounds-force
    2.343 845 602 . . . . . . . . . . . . . . . . . calories
    0.009 806 65 . . . . . . . . . . . . . . . . kilojoule*
    0.009 301 131 066 . . . . . . . . . . . . . . . . Btu
    0.002 724 069 444 . . . . . . . . . . . . . watt-hour

    0.002 343 845 602 . . . . . . . . . . . . . kilocalorie
    0.000 009 806 65 . . . . . . . . . . . . . megajoule*
    0.000 003 703 703 704 . . . . . . . metric horse-
                                              power-hour
    0.000 003 653 037 299 . . . . . . . . horsepower-
                                                  hour
    0.000 003 651 567 620 . . . . . . . electric horse-
                                              power-hour

    0.000 002 724 069 444 . . . . . . . kilowatt-hour
```

One KILOJOULE is equal to:

10 000 000 000 .... ergs*
1 000 000 000 .... microjoules*
10 197 162.13 .... gram-force centi-
                                          meters
23 730.360 40 .... foot poundals
1 000 .... joules*

737.562 149 3 .... foot pounds-force
239.005 736 1 .... calories
101.971 621 3 .... kilogram-force
                                      meters
0.948 451 414 8 .... Btu
0.277 777 777 8 .... watt-hour

0.239 005 736 1 .... kilocalorie
0.001 .... megajoule*
0.000 377 672 671 5 .... metric horse-
                                   power-hour
0.000 372 506 136 0 .... horsepower-
                                        hour
0.000 372 356 270 5 .... electric horse-
                                     power-hour

0.000 277 777 777 8 .... kilowatt-hour

One BTU is equal to:

1.054 350 264 × 10^{10} .... ergs
1.054 350 264 × 10^{9} .... microjoules
10 751 380.59 .... gram-force centi-
                                          meters
25 020.111 77 .... foot poundals
1 054.350 264 .... joules

777.648 847 2 .... foot pounds-force
251.995 761 1 .... calories
107.513 805 9 .... kilogram-force
                                      meters
1.054 350 264 .... kilojoules
0.292 875 073 5 .... watt-hour

0.251 995 761 1 . . . . kilocalorie
0.001 054 350 264 . . . . megajoule
0.000 398 199 281 1 . . . . metric horsepower-hour
0.000 392 751 943 0 . . . . horse-power-
hour
0.000 392 593 932 3 . . . . electric horsepower-hour
0.000 292 875 073 5 . . . . kilowatt-hour

One WATT-HOUR is equal to:

36 000 000 000 . . . . ergs*
3 600 000 000 . . . . microjoules*
36 709 783.67 . . . . gram-force centimeters
85 429.297 46 . . . . foot poundals
3 600 . . . . joules*

2 655.223 737 . . . . foot pounds-force
860.420 650 1 . . . . calories
367.097 836 7 . . . . kilogram-force
meters
3.6 . . . . kilojoules*
3.414 425 093 . . . . Btu

0.860 420 650 1 . . . . kilocalorie
0.003 6 . . . . megajoule*
0.001 359 621 617 . . . . metric horsepower-hour
0.001 341 022 090 . . . . horsepower-
hour
0.001 340 482 574 . . . . electric horsepower-hour
0.001 . . . . kilowatt-hour*

One KILOCALORIE is equal to:

41 840 000 000 . . . . ergs*
4 184 000 000 . . . . microjoules*

42 664 926.35 . . . . gram-force centimeters
99 287.827 93 . . . . foot poundals
4 184 . . . . joules*

3 085.960 033 . . . . foot pounds-force
1 000 . . . . calories*
426.649 263 5 . . . . kilogram-force
meters
4.184 . . . . kilojoules*
3.968 320 719 . . . . Btu

1.162 222 222 . . . . watt-hours
0.004 184 . . . . megajoule*
0.001 580 182 457 . . . . metric horsepower-hour
0.001 558 565 673 . . . . horsepower-
hour
0.001 557 938 636 . . . . electric horsepower-hour

0.001 162 222 222 . . . . kilowatt-hour

One MEGAJOULE is equal to:

10 000 000 000 000 . . . . ergs*
1 000 000 000 000 . . . . microjoules*
1.019 716 213 × 10^{10} . . . . . . gram-force centi-
meters
23 730 360.40 . . . . foot poundals
1 000 000 . . . . joules*

737 562.149 3 . . . . foot pounds-force
239 005.736 1 . . . . calories
101 971.621 3 . . . . kilogram-force
meters
1 000 . . . . kilojoules*
948.451 414 8 . . . . Btu

277.777 777 8 . . . . watt-hours
239.005 736 1 . . . . kilocalories
0.377 672 671 5 . . . . metric horsepower-hour

0.372 506 136 0 . . . . horsepower-
hour
0.372 356 270 5 . . . . electric horsepower-hour

0.277 777 777 8 . . . . kilowatt-hour

One **METRIC HORSEPOWER-HOUR** is equal to:

26 477 955 000 000 . . . . ergs*
2 647 795 500 000 . . . . microjoules*
27 000 000 000 . . . . gram-force centimeters*
62 833 141.49 . . . . foot poundals
2 647 795.5 . . . . joules*

1 952 913.740 . . . . foot pounds-force
632 838.312 6 . . . . calories
270 000 . . . . kilogram-force
meters*
2 647.795 5 . . . . kilojoules*
2 511.305 388 . . . . Btu

735.498 75 . . . . watt-hours*
632.838 312 6 . . . . kilocalories
2.647 795 5 . . . . megajoules*
0.986 320 070 6 . . . . horsepower-
hour
0.985 923 257 4 . . . . electric horsepower-hour

0.735 498 75 . . . . kilowatt-hour*

One **HORSEPOWER-HOUR** is equal to:

26 845 195 376 961.727 92 . . . . ergs*
2 684 519 537 696.172 792 . . . . microjoules*
27 374 480 966.448 . . . . gram-force centimeters*
63 704 616.14 . . . . foot poundals
2 684 519.537 696 172 792 . . . . joules*

```
1 980 000 . . . . . . . . . . . . . . . . . . . . foot pounds-force*
  641 615.568 3 . . . . . . . . . . . . . . . . . . . . . . . . calories
  273 744.809 664 48 . . . . . . . . . . . . . kilogram-force
                                                       meters*
    2 684.519 537 696 172 792 . . . . . . . . . kilojoules*
    2 546.136 353 . . . . . . . . . . . . . . . . . . . . . . . . Btu

     745.699 871 582 270 22 . . . . . . . . . watt-hours*
     641.615 568 3 . . . . . . . . . . . . . . . . . kilocalories
       2.684 519 537 696 172 792 . . . . . megajoules*
       1.013 869 665 424 . . . . . . . . . . metric horse-
                                                       power-hours*
       0.999 597 683 1 . . . . . . . . . . . . electric horse-
                                                       power-hour

       0.745 699 871 582 270 22 . . . . kilowatt-hour*
```

One ELECTRIC HORSEPOWER-HOUR is equal to:

```
26 856 000 000 000 . . . . . . . . . . . . . . . . . . . . . . . . . . . . . . ergs*
 2 685 600 000 000 . . . . . . . . . . . . . . . . . . . . . . . . . microjoules*
             2.738 549 862 × 10^{10} . . . . . . gram-force centi-
                                                          meters
   63 730 255.90 . . . . . . . . . . . . . . . . . . . . . foot poundals
    2 685 600 . . . . . . . . . . . . . . . . . . . . . . . . . . . joules*

    1 980 796.908 . . . . . . . . . . . . . . . . foot pounds-force
      641 873.805 0 . . . . . . . . . . . . . . . . . . . . . . . . calories
      273 854.986 2 . . . . . . . . . . . . . . . . . . kilogram-force
                                                          meters
        2 685.6 . . . . . . . . . . . . . . . . . . . . . . . . kilojoules*
        2 547.161 119 . . . . . . . . . . . . . . . . . . . . . . . Btu

        746 . . . . . . . . . . . . . . . . . . . . . . . . watt-hours*
        641.873 805 0 . . . . . . . . . . . . . . . . . kilocalories
          2.685 6 . . . . . . . . . . . . . . . . . . . . . megajoules*
          1.014 277 727 . . . . . . . . . . . . . metric horse-
                                                          power-hours
          1.000 402 479 . . . . . . . . . . . . . horsepower-
                                                          hours

          0.746 . . . . . . . . . . . . . . . . . . . . kilowatt-hour*
```

One **KILOWATT-HOUR** is equal to:

36 000 000 000 000 . . . . ergs*

3 600 000 000 000 . . . . microjoules*

3.670 978 367 × 10^{10} . . . . . . gram-force centimeters

85 429 297.46 . . . . foot poundals

3 600 000 . . . . joules*

2 655 223.737 . . . . foot pounds-force

860 420.650 1 . . . . calories

367 097.836 7 . . . . kilogram-force
meters

3 600 . . . . kilojoules*

3 414.425 093 . . . . Btu

1 000 . . . . watt-hours*

860.420 650 1 . . . . kilocalories

3.6 . . . . megajoules*

1.359 621 617 . . . . metric horsepower-hours

1.341 022 090 . . . . horsepower-
hours

1.340 482 574 . . . . electric horsepower-hours

ADDITIONAL UNITS OF ENERGY

Metric

1 electronvolt =	1.602 189 2 × 10^{-19} . . . . joule	
1 dyne-centi-		
meter =	1 . . . . erg* (in tables)	
=	0.000 000 1 . . . . joule*	
1 cubic centi-		
meter-atmo-		
sphere =	0.101 325 . . . . joule*	
1 newton-		
meter =	1 . . . . joule* (in tables)	

1 volt-
   coulomb   =   1 . . . . joule* (in tables)
1 watt-
   second   =   1 . . . . joule* (in tables)

1 gram calorie
   (20°C)   =   4.181 90 . . . . joules
1 thermochemi-
   cal gram
   calorie   =   1 . . . . calorie* (in tables)
              =   4.184 . . . . joules*
1 gram calorie
   (15°C)   =   4.185 80 . . . . joules
1 International
   Steam Table
1 gram calorie =   4.186 8 . . . . joules*
1 mean gram
   calorie   =   4.190 02 . . . . joules

1 liter-atmo-
   sphere   =   101.325 . . . . joules*

1 thermochemi-
   cal kilogram
   calorie   =   1 . . . . kilocalorie* (in tables)
              =   4 184 . . . . joules*

1 International
   Steam Table
1 kilogram
   calorie   =   4 186.8 . . . . joules*
1 mean
   kilogram
   calorie   =   4 190.02 . . . . joules

1 cheval-
   vapeur-
   heure   =   1 . . . . metric horse-
                                                              power-hour*
                                                              (in tables)
              = 2 647 795.5 . . . . joules*

Customary

1 thermochemi- cal Btu	=	1 . . . . Btu* (in tables)
	=	1 054.350 . . . . joules
1 Btu (60°F)	=	1 054.68 . . . . joules
1 Btu (59°F)	=	1 054.80 . . . . joules
1 International Steam Table Btu	=	1 055.056 . . . . joules
1 mean Btu	=	1 055.87 . . . . joules
1 Btu (39°F)	=	1 059.67 . . . . joules
1 cubic foot- atmosphere	=	2 869.204 . . . . joules
1 therm	=	100 000 . . . . International Steam Table Btu*
	=	$1.055\ 056 \times 10^8$ . . . . joules
nuclear equiva- lent of 1 ton of TNT	= 1 000 000 000 . . . . calories*	
	= 4 184 000 000 . . . . joules*	

Flow

One CUBIC CENTIMETER or MILLILITER PER SECOND is equal to:

0.06 . . . . cubic decimeter or liter per minute*

0.022 643 318 77 . . . . petroleum barrel per hour

0.015 850 323 14 . . . . gallon per minute

0.002 118 880 003 . . . . cubic foot per minute

0.001 . . . . cubic decimeter or liter per second*

0.000 264 172 052 4 . . . . gallon per second

0.000 078 477 037 16 . . . . cubic yard per minute

0.000 06 . . . . cubic meter or kiloliter per minute*

0.000 035 314 666 72 . . . . cubic foot per second

0.000 001 . . . . cubic meter or kiloliter per second*

One CUBIC DECIMETER or LITER PER MINUTE is equal to:

16.666 666 67 . . . . cubic centimeters or milliliters/second

0.377 388 646 2 . . . . petroleum barrel per hour

0.264 172 052 4 . . . . gallon per minute

0.035 314 666 72 . . . . cubic foot per minute

0.016 666 666 67 . . . cubic decimeter or liter per second

0.004 402 867 539 . . . . gallon per second

0.001 307 950 619 . . . . . cubic yard per minute

109

0.001 . . . . cubic meter or kiloliter
per minute*
0.000 588 577 778 7 . . . . cubic foot per second
0.000 016 666 666 67 . . . . cubic meter or kilo-
liter/second

One **PETROLEUM BARREL PER HOUR** is equal to:

44.163 137 48 . . . . cubic centimeters or milli-
liters/second*
2.649 788 248 8 . . . . . cubic decimeters or liters
per minute*
0.7 . . . . gallon per minute*
0.093 576 388 89 . . . . cubic foot per minute
0.044 163 137 48 . . . . . . cubic decimeter or liter
per second*

0.011 666 666 67 . . . . gallon per second
0.003 465 792 181 . . . . . cubic yard per minute
0.002 649 788 248 8 . . . . . cubic meter or kilo-
liter/minute*
0.001 559 606 481 . . . . . . cubic foot per second
0.000 044 163 137 48 . . . . cubic meter or kilo-
liter/second*

One **GALLON PER MINUTE** is equal to:

63.090 196 4 . . . . cubic centimeters or milli-
liters/second*
3.785 411 784 . . . . . . cubic decimeters or liters
per minute*
1.428 571 429 . . . . . petroleum barrels per hour
0.133 680 555 6 . . . . cubic foot per minute
0.063 090 196 4 . . . . cubic decimeter or liter
per second*

0.016 666 666 67 . . . . gallon per second
0.004 951 131 687 . . . . . cubic yard per minute
0.003 785 411 784 . . . . . cubic meter or kiloliter
per minute*
0.002 228 009 259 . . . . . . cubic foot per second
0.000 063 090 196 4 . . cubic meter or kiloliter/
second*

One CUBIC FOOT PER MINUTE is equal to:

471.947 443 2 . . . . cubic centimeters or milli-
liters/second*
28.316 846 592 . . . . . . cubic decimeters or liters
per minute*
10.686 456 40 . . . . . petroleum barrels per hour
7.480 519 481 . . . . gallons per minute
0.471 947 443 2 . . . . cubic decimeter or liter
per second*

0.124 675 324 7 . . . . gallon per second
0.037 037 037 04 . . . . . . cubic yard per minute
0.028 316 846 592 . . . . . cubic meter or kiloliter
per minute*
0.016 666 666 67 . . . . cubic foot per second
0.000 471 947 443 2 . . cubic meter or kiloliter/
second*

One CUBIC DECIMETER or LITER PER SECOND is equal to:

1 000 . . . . cubic centimeters or milliliters
per second*
60 . . . . cubic decimeters or liters
per minute*
22.643 318 77 . . . . . . petroleum barrels per hour
15.850 323 14 . . . . gallons per minute
2.118 880 003 . . . . cubic feet per minute

0.264 172 052 4 . . . . gallon per second
0.078 477 037 16 . . . . . . cubic yard per minute
0.06 . . . . cubic meter or kiloliter
per minute*
0.035 314 666 72 . . . . cubic foot per second
0.001 . . . . cubic meter or kiloliter
per second*

One GALLON PER SECOND is equal to:

3 785.411 784 . . . . cubic centimeters or milli-
liters/second*
227.124 707 04 . . . . cubic decimeters or liters
per minute*

85.714 285 71 . . . . petroleum barrels per
hour
60 . . . . gallons per minute*
8.020 833 333 . . . . cubic feet per minute
3.785 411 784 . . . . . . cubic decimeters or liters
per second*
0.297 067 901 2 . . . . . . cubic yard per minute
0.227 124 707 04 . . . . . . cubic meter or kiloliter
per minute*
0.133 680 555 6 . . . . cubic foot per second
0.003 785 411 784 . . . . . cubic meter or kiloliter
per second*

One CUBIC YARD PER MINUTE is equal to:

12 742.580 966 4 . . . . cubic centimeters or milli-
liters/second*
764.554 857 984 . . . . . . cubic decimeters or liters
per minute*
288.534 322 8 . . . . . . petroleum barrels per hour
201.974 026 0 . . . . gallons per minute
27 . . . . cubic feet per minute*
12.742 580 966 4 . . . . . cubic decimeters or liters
per second*
3.366 233 766 . . . . gallons per second
0.764 554 857 984 . . . . . cubic meter or kiloliter
per minute*
0.45 . . . . cubic foot per second*
0.012 742 580 966 4 . . . . . cubic meter or kilo-
liter/second*

One CUBIC METER or KILOLITER PER MINUTE is equal to:

16 666.666 67 . . . . cubic centimeters or milliliters per second
1 000 . . . . cubic decimeters or liters
per minute*
377.388 646 2 . . . . . . petroleum barrels per hour
264.172 052 4 . . . . gallons per minute

35.314 666 72 . . . . cubic feet per minute

16.666 666 67 . . . . cubic decimeters or liters
per second

4.402 867 539 . . . . gallons per second

1.307 950 619 . . . . cubic yards per minute

0.588 577 778 7 . . . . cubic foot per second

0.016 666 666 67 . . . . . . cubic meter or kiloliter
per second

One CUBIC FOOT PER SECOND is equal to:

28 316.846 592 . . . . cubic centimeters or milli-
liters/second*

1 699.010 795 52 . . . . . . cubic decimeters or liters
per minute*

641.187 384 0 . . . . . . petroleum barrels per hour

448.831 168 8 . . . . gallons per minute

60 . . . . cubic feet per minute*

28.316 846 592 . . . . . . cubic decimeters or liters
per second*

7.480 519 481 . . . . gallons per second

2.222 222 222 . . . . cubic yards per minute

1.699 010 795 52 . . . . cubic meters or kilo-
liters/minute*

0.028 316 846 592 . . . . cubic meter or kiloliter per second*

One CUBIC METER or KILOLITER PER SECOND is equal to:

1 000 000 . . . . cubic centimeters or milliliters per second*

60 000 . . . . cubic decimeters or liters
per minute*

22 643.318 77 . . . . petroleum barrels per hour

15 850.323 14 . . . . gallons per minute

2 118.880 003 . . . . cubic feet per minute

1 000 . . . . cubic decimeters or liters
per second*

$$264.172\ 052\ 4 \ldots\ldots\ldots\ldots \text{ gallons per second}$$
$$78.477\ 037\ 16 \ldots\ldots\ldots \text{ cubic yards per minute}$$
$$60 \ldots\ldots\ldots\ldots \text{ cubic meters or kiloliters per minute*}$$
$$35.314\ 666\ 72 \ldots\ldots\ldots \text{ cubic feet per second}$$

ADDITIONAL UNITS OF FLOW

Metric

1 cubic centimeter or milliliter/day	=	$1.157\ 407 \times 10^{-8}$ . . . . . liter per second
1 cubic centimeter or milliliter/hour	=	$2.777\ 778 \times 10^{-7}$ . . . . . liter per second
1 cubic decimeter or liter per day	=	0.000 011 574 07 . . . . . liter per second
1 cubic centimeter or milliliter/minute	=	0.000 016 666 67 . . . . . liter per second
1 cubic decimeter or liter per hour	=	0.000 277 777 8 . . . . . . liter per second
1 cubic meter or kiloliter per day	=	0.011 574 07 . . . . liter per second
1 cubic meter or kiloliter per hour	=	0.277 777 8 . . . . liter per second

Customary

1 cubic inch per day	=	$1.896\ 651 \times 10^{-7}$ . . . . . liter per second
1 cubic inch per hour	=	0.000 004 551 962 . . . . liter per second
1 gallon per day	=	0.000 043 812 64 . . . . . liter per second
1 cubic inch per minute	=	0.000 273 117 7 . . . . . . liter per second
1 cubic foot per day	=	0.000 327 741 28 . . . . . . liter/second*
1 gallon per hour	=	0.001 051 503 . . . . liter per second
1 petroleum barrel per day	=	0.001 840 131 . . . . liter per second
1 cubic foot per hour	=	0.007 865 791 . . . . liter per second
1 cubic yard per day	=	0.008 849 015 . . . . liter per second
1 cubic inch per second	=	0.016 387 064 . . . . liter per second*

1 cubic yard per hour   =   0.212 376 3 . . . . liter per second
1 petroleum barrel per
   minute   =   2.649 788 . . . . liters per second
1 petroleum barrel per
   second   = 158.987 294 928 . . . . liters/second*
1 cubic yard per second = 764.554 857 984 . . . . liters/second*

Force

One DYNE is equal to:

    0.001 019 716 213 . . . . gram-force
    0.000 072 330 138 51 . . . . poundal
    0.000 01 . . . . newton*
    0.000 002 248 089 431 . . . . pound-force
    0.000 001 019 716 213 . . . . kilogram-force

    0.000 000 01 . . . . kilonewton*
    1.124 044 715 $\times$ 10^{-9} . . . . short ton-force
    1.019 716 213 $\times$ 10^{-9} . . . . . . metric ton-force

One GRAM-FORCE is equal to:

    980.665 . . . . dynes*
    0.070 931 635 28 . . . . poundal
    0.009 806 65 . . . . newton*
    0.002 204 622 622 . . . . pound-force
    0.001 . . . . kilogram-force*

    0.000 009 806 65 . . . . kilonewton*
    0.000 001 102 311 311 . . . . short ton-force
    0.000 001 . . . . metric ton-force*

One POUNDAL is equal to:

    13 825.495 437 6 . . . . dynes*
    14.098 081 85 . . . . grams-force
    0.138 254 954 376 . . . . newton*
    0.031 080 950 17 . . . . pound-force
    0.014 098 081 85 . . . . kilogram-force

    0.000 138 254 954 376 . . . . kilonewton*
    0.000 015 540 475 09 . . . . short ton-force
    0.000 014 098 081 85 . . . . . . metric ton-force

One NEWTON is equal to:

```
100 000 . . . . . . . . . . . . . . . . . . . . . . . . . . . . . dynes*
    101.971 621 3 . . . . . . . . . . . . . . . . grams-force
      7.233 013 851 . . . . . . . . . . . . . . . poundals
      0.224 808 943 1 . . . . . . . . . . . . pound-force
      0.101 971 621 3 . . . . . . . . . . . kilogram-force

      0.001 . . . . . . . . . . . . . . . . . . . . . . kilonewton*
      0.000 112 404 471 5 . . . . . . . . short ton-force
      0.000 101 971 621 3 . . . . . . . metric ton-force
```

One POUND-FORCE is equal to:

```
444 822.161 526 05 . . . . . . . . . . . . . . . . . . dynes*
    453.592 37 . . . . . . . . . . . . . . . . . . . . grams-force*
     32.174 048 56 . . . . . . . . . . . . . . . . poundals
      4.448 221 615 260 5 . . . . . . . . . . . newtons*
      0.453 592 37 . . . . . . . . . . . . . kilogram-force*

      0.004 448 221 615 260 5 . . . . . . . kilonewton*
      0.000 5 . . . . . . . . . . . . . . . . . . short ton-force*
      0.000 453 592 37 . . . . . . . . . metric ton-force*
```

One KILOGRAM-FORCE is equal to:

```
980 665 . . . . . . . . . . . . . . . . . . . . . . . . . . . . . dynes*
  1 000 . . . . . . . . . . . . . . . . . . . . . . . . . . . . grams-force*
     70.931 635 28 . . . . . . . . . . . . . . . . . poundals
      9.806 65 . . . . . . . . . . . . . . . . . . . . . . newtons*
      2.204 622 622 . . . . . . . . . . . . . . pounds-force

      0.009 806 65 . . . . . . . . . . . . . . . kilonewton*
      0.001 102 311 311 . . . . . . . . . . short ton-force
      0.001 . . . . . . . . . . . . . . . . . metric ton-force*
```

One KILONEWTON is equal to:

```
100 000 000 . . . . . . . . . . . . . . . . . . . . . . . . . . . dynes*
    101 971.621 3 . . . . . . . . . . . . . . . . . . . . grams-force
      7 233.013 851 . . . . . . . . . . . . . . . . . . . poundals
      1 000 . . . . . . . . . . . . . . . . . . . . . . . . . . . newtons*
```

224.808 943 1 . . . . pounds-force

101.971 621 3 . . . . kilograms-force
0.112 404 471 5 . . . . short ton-force
0.101 971 621 3 . . . . metric ton-force

One SHORT TON-FORCE is equal to:

889 644 323.052 1 . . . . dynes*
907 184.74 . . . . grams-force*
64 348.097 11 . . . . poundals
8 896.443 230 521 . . . . newtons*
2 000 . . . . pounds-force*

907.184 74 . . . . kilograms-force*
8.896 443 230 521 . . . . kilonewtons*
0.907 184 74 . . . . metric ton-force*

One METRIC TON-FORCE is equal to:

980 665 000 . . . . dynes*
1 000 000 . . . . grams-force*
70 931.635 28 . . . . poundals
9 806.65 . . . . newtons*
2 204.622 622 . . . . pounds-force

1 000 . . . . kilograms-force*
9.806 65 . . . . kilonewtons*
1.102 311 311 . . . . short tons-force

ADDITIONAL UNITS OF FORCE

Metric

1 kilopound-force =   1 . . . . kilogram-force* (in
tables)
=   9.806 65 . . . . newtons*

Customary

1 ounce-force   =   0.278 013 9 . . . . newton
1 kip   = 1 000 . . . . pounds-force*
= 4 448.221 615 260 5 . . . . newtons*

Length

One ANGSTROM is equal to:

0.000 1 . . . . micrometer*
0.000 000 1 . . . . millimeter*
0.000 000 01 . . . . centimeter*
$3.937\ 007\ 874 \times 10^{-9}$ . . . . inch
$4.970\ 959\ 596 \times 10^{-10}$ . . . . link

$3.280\ 839\ 895 \times 10^{-10}$ . . . . foot
$3.280\ 833\ 333 \times 10^{-10}$ . . . . survey foot
$1.093\ 613\ 298 \times 10^{-10}$ . . . . yard
0.000 000 000 1 . . . . meter*
$1.988\ 383\ 838 \times 10^{-11}$ . . . . rod

$4.970\ 959\ 596 \times 10^{-12}$ . . . . chain
0.000 000 000 000 1 . . . . kilometer*
$6.213\ 711\ 922 \times 10^{-14}$ . . . . mile
$6.213\ 699\ 495 \times 10^{-14}$ . . . . survey mile
$5.399\ 568\ 035 \times 10^{-14}$ . . . . nautical mile

One MICROMETER is equal to:

10 000 . . . . angstroms*
0.001 . . . . millimeter*
0.000 1 . . . . centimeter*
0.000 039 370 078 74 . . . . inch
0.000 004 970 959 596 . . . . link

0.000 003 280 839 895 . . . . foot
0.000 003 280 833 333 . . . . survey foot
0.000 001 093 613 298 . . . . yard
0.000 001 . . . . meter*
$1.988\ 383\ 838 \times 10^{-7}$ . . . . rod

$4.970\ 959\ 596 \times 10^{-8}$ . . . . chain

```
0.000 000 001 . . . . . . . . . . . . . . . kilometer*
6.213 711 922 × 10⁻¹⁰ . . . . . . . . . . . . . . mile
6.213 699 495 × 10⁻¹⁰ . . . . . . . . . . survey mile
5.399 568 035 × 10⁻¹⁰ . . . . . . . . nautical mile
```

One MILLIMETER is equal to:

```
10 000 000 . . . . . . . . . . . . . . . . . . . . . . . . . . . angstroms*
    1 000 . . . . . . . . . . . . . . . . . . . . . . . micrometers*
      0.1 . . . . . . . . . . . . . . . . . . . . . . . . centimeter*
      0.039 370 078 74 . . . . . . . . . . . . . . . . . inch
      0.004 970 959 596 . . . . . . . . . . . . . . . . . link

      0.003 280 839 895 . . . . . . . . . . . . . . . . foot
      0.003 280 833 333 . . . . . . . . . . . . . survey foot
      0.001 093 613 298 . . . . . . . . . . . . . . . yard
      0.001 . . . . . . . . . . . . . . . . . . . . . . . . meter*
      0.000 198 838 383 8 . . . . . . . . . . . . . . . rod

      0.000 049 709 595 96 . . . . . . . . . . . . . chain
      0.000 001 . . . . . . . . . . . . . . . . . . kilometer*
      6.213 711 922 × 10⁻⁷ . . . . . . . . . . . . . . mile
      6.213 699 495 × 10⁻⁷ . . . . . . . . . . survey mile
      5.399 568 035 × 10⁻⁷ . . . . . . . . nautical mile
```

One CENTIMETER is equal to:

```
100 000 000 . . . . . . . . . . . . . . . . . . . . . . . . . angstroms*
 10 000 . . . . . . . . . . . . . . . . . . . . . . . micrometers*
     10 . . . . . . . . . . . . . . . . . . . . . . millimeters*
      0.393 700 787 4 . . . . . . . . . . . . . . . . . inch
      0.049 709 595 96 . . . . . . . . . . . . . . . . . link

      0.032 808 398 95 . . . . . . . . . . . . . . . . foot
      0.032 808 333 33 . . . . . . . . . . . . . survey foot
      0.010 936 132 98 . . . . . . . . . . . . . . . yard
      0.01 . . . . . . . . . . . . . . . . . . . . . . . . meter*
      0.001 988 383 838 . . . . . . . . . . . . . . . rod

      0.000 497 095 959 6 . . . . . . . . . . . . . chain
      0.000 01 . . . . . . . . . . . . . . . . . . kilometer*
```

```
0.000  006  213  711  922 . . . . . . . . . . . . . . mile
0.000  006  213  699  495 . . . . . . . . . . survey mile
0.000  005  399  568  035 . . . . . . . . . nautical mile
```

One INCH is equal to:

```
254  000  000 . . . . . . . . . . . . . . . . . . . . . . . . . . . angstroms*
     25  400 . . . . . . . . . . . . . . . . . . . . . . . . micrometers*
        25.4 . . . . . . . . . . . . . . . . . . . . . . . . millimeters*
        2.54 . . . . . . . . . . . . . . . . . . . . . . . centimeters*
0.126  262  373  7 . . . . . . . . . . . . . . . . . . . link

0.083  333  333  33 . . . . . . . . . . . . . . . . foot
0.083  333  166  67 . . . . . . . . . . . . . . survey foot
0.027  777  777  78 . . . . . . . . . . . . . . . . yard
0.025  4 . . . . . . . . . . . . . . . . . . . . . . . . meter*
0.005  050  494  949 . . . . . . . . . . . . . . . . rod

0.001  262  623  737 . . . . . . . . . . . . . . . . chain
0.000  025  4 . . . . . . . . . . . . . . . . . . kilometer*
0.000  015  782  828  28 . . . . . . . . . . . . . mile
0.000  015  782  796  72 . . . . . . . . . . survey mile
0.000  013  714  902  81 . . . . . . . . . nautical mile
```

One LINK is equal to:

```
2.011  684  023  ×  10^9 . . . . . . . . . . . angstroms
201  168.402  3 . . . . . . . . . . . . . . . . . . . . micrometers
201.168  402  3 . . . . . . . . . . . . . . . . . millimeters
20.116  840  23 . . . . . . . . . . . . . . . centimeters
7.920  015  840 . . . . . . . . . . . . . . . . . . inches

0.660  001  320  0 . . . . . . . . . . . . . . . . . . foot
0.66 . . . . . . . . . . . . . . . . . . . . . . survey foot*
0.220  000  440  0 . . . . . . . . . . . . . . . . . yard
0.201  168  402  3 . . . . . . . . . . . . . . . . meter
0.04 . . . . . . . . . . . . . . . . . . . . . . . . . . . rod*

0.01 . . . . . . . . . . . . . . . . . . . . . . . . . . . chain*
0.000  201  168  402  3 . . . . . . . . . . . . . kilometer
0.000  125  000  250  0 . . . . . . . . . . . . . . mile
0.000  125 . . . . . . . . . . . . . . . . . . . . survey mile*
0.000  108  622  247  5 . . . . . . . . . . nautical mile
```

One FOOT is equal to:

3 048 000 000 . . . . angstroms*
304 800 . . . . micrometers*
304.8 . . . . millimeters*
30.48 . . . . centimeters*
12 . . . . inches*

1.515 148 485 . . . . links
0.999 998 . . . . survey foot*
0.333 333 333 3 . . . . yard
0.304 8 . . . . meter*
0.060 605 939 39 . . . . rod

0.015 151 484 85 . . . . chain
0.000 304 8 . . . . kilometer*
0.000 189 393 939 4 . . . . mile
0.000 189 393 560 6 . . . . survey mile
0.000 164 578 833 7 . . . . nautical mile

One SURVEY FOOT is equal to:

3.048 006 096 × 10^9 . . . . angstroms
304 800.609 6 . . . . micrometers
304.800 609 6 . . . . millimeters
30.480 060 96 . . . . centimeters
12.000 024 00 . . . . inches

1.515 151 515 . . . . links
1.000 002 000 . . . . feet
0.333 334 000 0 . . . . yard
0.304 800 609 6 . . . . meter
0.060 606 060 61 . . . . rod

0.015 151 515 15 . . . . chain
0.000 304 800 609 6 . . . . kilometer
0.000 189 394 318 2 . . . . mile
0.000 189 393 939 4 . . . . survey mile
0.000 164 579 162 9 . . . . nautical mile

One YARD is equal to:

```
9 144 000 000  . . . . . . . . . . . . . . . . . . . . . . . . . angstroms*
      914 400  . . . . . . . . . . . . . . . . . . . . . micrometers*
        914.4  . . . . . . . . . . . . . . . . . . . . . millimeters*
        91.44  . . . . . . . . . . . . . . . . . . . . centimeters*
           36  . . . . . . . . . . . . . . . . . . . . . . . inches*

    4.545 445 455  . . . . . . . . . . . . . . . . . . . links
    3  . . . . . . . . . . . . . . . . . . . . . . . . . . . . feet*
    2.999 994  . . . . . . . . . . . . . . . . . . survey feet*
    0.914 4 . . . . . . . . . . . . . . . . . . . . . . meter*
    0.181 817 818 2 . . . . . . . . . . . . . . . . . . rod

    0.045 454 454 55  . . . . . . . . . . . . . . . . chain
    0.000 914 4  . . . . . . . . . . . . . . . . kilometer*
    0.000 568 181 818 2  . . . . . . . . . . . . . . mile
    0.000 568 180 681 8  . . . . . . . . . . survey mile
    0.000 493 736 501 1  . . . . . . . . . nautical mile
```

One METER is equal to:

```
10 000 000 000  . . . . . . . . . . . . . . . . . . . . . . . angstroms*
    1 000 000  . . . . . . . . . . . . . . . . . . . . . micrometers*
        1 000  . . . . . . . . . . . . . . . . . . . . . millimeters*
          100  . . . . . . . . . . . . . . . . . . . . centimeters*
   39.370 078 74  . . . . . . . . . . . . . . . . . . inches

    4.970 959 596  . . . . . . . . . . . . . . . . . . links
    3.280 839 895  . . . . . . . . . . . . . . . . . . feet
    3.280 833 333  . . . . . . . . . . . . . . survey feet
    1.093 613 298  . . . . . . . . . . . . . . . . . yards
    0.198 838 383 8 . . . . . . . . . . . . . . . . . . rod

    0.049 709 595 96  . . . . . . . . . . . . . . . . chain
    0.001  . . . . . . . . . . . . . . . . . . . . . kilometer*
    0.000 621 371 192 2  . . . . . . . . . . . . . . mile
    0.000 621 369 949 5  . . . . . . . . . . survey mile
    0.000 539 956 803 5  . . . . . . . . . nautical mile
```

One ROD is equal to:

$$5.029\ 210\ 058 \times 10^{10} \ \ldots\ldots\ldots\ \text{angstroms}$$

5 029 210.058 . . . . micrometers
5 029.210 058 . . . . millimeters
502.921 005 8 . . . . centimeters
198.000 396 0 . . . . inches

25 . . . . links*
16.500 033 00 . . . . feet
16.5 . . . . survey feet*
5.500 011 000 . . . . yards
5.029 210 058 . . . . meters

0.25 . . . . chain*
0.005 029 210 058 . . . . kilometer
0.003 125 006 250 . . . . mile
0.003 125 . . . . survey mile*
0.002 715 556 187 . . . . nautical mile

One CHAIN is equal to:

$$2.011\ 684\ 023 \times 10^{11} \ \ldots\ldots\ldots\ \text{angstroms}$$

20 116 840.23 . . . . micrometers
20 116.840 23 . . . . millimeters
2 011.684 023 . . . . centimeters
792.001 584 0 . . . . inches

100 . . . . links*
66.000 132 00 . . . . feet
66 . . . . survey feet*
22.000 044 00 . . . . yards
20.116 840 23 . . . . meters

4 . . . . rods*
0.020 116 840 23 . . . . kilometer
0.012 500 025 00 . . . . mile
0.012 5 . . . . survey mile*
0.010 862 224 75 . . . . nautical mile

One KILOMETER is equal to:

```
10 000 000 000 000 . . . . . . . . . . . . . . . . . . . . . . . . . . . angstroms*
     1 000 000 000 . . . . . . . . . . . . . . . . . . . . . . . . micrometers*
         1 000 000 . . . . . . . . . . . . . . . . . . . . . . . millimeters*
           100 000 . . . . . . . . . . . . . . . . . . . . . . centimeters*
        39 370.078 74 . . . . . . . . . . . . . . . . . . . . . . . inches

         4 970.959 596 . . . . . . . . . . . . . . . . . . . . . . . links
         3 280.839 895 . . . . . . . . . . . . . . . . . . . . . . . feet
         3 280.833 333 . . . . . . . . . . . . . . . . . . . survey feet
         1 093.613 298 . . . . . . . . . . . . . . . . . . . . . . yards
         1 000 . . . . . . . . . . . . . . . . . . . . . . . . . . . meters*

           198.838 383 8 . . . . . . . . . . . . . . . . . . . . rods
            49.709 595 96 . . . . . . . . . . . . . . . . . . . chains
             0.621 371 192 2 . . . . . . . . . . . . . . . . . . mile
             0.621 369 949 5 . . . . . . . . . . . . . . survey mile
             0.539 956 803 5 . . . . . . . . . . . . . nautical mile
```

One MILE is equal to:

```
16 093 440 000 000 . . . . . . . . . . . . . . . . . . . . . . . . angstroms*
     1 609 344 000 . . . . . . . . . . . . . . . . . . . . . . micrometers*
         1 609 344 . . . . . . . . . . . . . . . . . . . . . . millimeters*
           160 934.4 . . . . . . . . . . . . . . . . . . . . . centimeters*
            63 360 . . . . . . . . . . . . . . . . . . . . . . . inches*

         7 999.984 . . . . . . . . . . . . . . . . . . . . . . . links*
         5 280 . . . . . . . . . . . . . . . . . . . . . . . . . feet*
         5 279.989 44 . . . . . . . . . . . . . . . . . . . survey feet*
         1 760 . . . . . . . . . . . . . . . . . . . . . . . . . yards*
         1 609.344 . . . . . . . . . . . . . . . . . . . . . . . meters*

           319.999 36 . . . . . . . . . . . . . . . . . . . . . rods*
            79.999 84 . . . . . . . . . . . . . . . . . . . . chains*
             1.609 344 . . . . . . . . . . . . . . . . . . kilometers*
             0.999 998 . . . . . . . . . . . . . . . . . survey mile*
             0.868 976 241 9 . . . . . . . . . . . . . nautical mile
```

One SURVEY MILE is equal to:

$$1.609\ 347\ 219 \times 10^{13} \quad \ldots\ldots\ldots \text{ angstroms}$$
$$1.609\ 347\ 219 \times 10^{9} \quad \ldots\ldots\ldots \text{ micrometers}$$

1 609 347.219 . . . . millimeters
160 934.721 9 . . . . centimeters
63 360.126 72 . . . . inches

8 000 . . . . links*
5 280.010 560 . . . . feet
5 280 . . . . survey feet*
1 760.003 520 . . . . yards
1 609.347 219 . . . . meters

320 . . . . rods*
80 . . . . chains*
1.609 347 219 . . . . kilometers
1.000 002 000 . . . . miles
0.868 977 979 9 . . . . nautical mile

One NAUTICAL MILE is equal to:

18 520 000 000 000 . . . . angstroms*
1 852 000 000 . . . . micrometers*
1 852 000 . . . . millimeters*
185 200 . . . . centimeters*
72 913.385 83 . . . . inches

9 206.217 172 . . . . links
6 076.115 486 . . . . feet
6 076.103 333 . . . . survey feet
2 025.371 829 . . . . yards
1 852 . . . . meters*

368.248 686 9 . . . . rods
92.062 171 72 . . . . chains
1.852 . . . . kilometers*
1.150 779 448 . . . . miles
1.150 777 146 . . . . survey miles

ADDITIONAL UNITS OF LENGTH

Metric

1 femtometer or fermi	=	1×10^{-15}	. . . . meter*
1 nanometer	=	1×10^{-9}	. . . . meter*
1 micron	=	1	. . . . micrometer* (in tables)
1 decimeter	=	0.1	. . . . meter*
1 dekameter	=	10	. . . . meters*
1 hectometer	=	100	. . . . meters*
1 geographical mile	=	1	. . . . nautical mile* (in tables)
1 international nautical mile	=	1	. . . . nautical mile* (in tables)
1 sea mile	=	1	. . . . nautical mile* (in tables)
1 U.S. nautical mile	=	1	. . . . nautical mile* (in tables)
1 international nautical league	=	3	. . . . nautical miles*
1 gigameter	=	1×10^{9}	. . . . meters*

Customary

1 mil	=	0.001	. . . . inch*
	=	0.025 4	. . . . millimeter*
1 caliber	=	0.01	. . . . inch*
	=	0.254	. . . . millimeter*
1 printer's point	=	0.013 837	. . . . inch*
	=	0.351 459 8	. . . . millimeter*
1 typography point	=	0.013 837	. . . . inch*
	=	0.351 459 8	. . . . millimeter*
1 printer's pica	=	0.166 044	. . . . inch*
	=	4.217 517 6	. . . . millimeters
1 palm	=	3	. . . . inches*
	=	7.62	. . . . centimeters*
1 hand	=	4	. . . . inches*
	=	10.16	. . . . centimeters*
1 Gunter's link	=	1	. . . . link* (in tables)
1 surveyor's link	=	1	. . . . link* (in tables)
1 span	=	9	. . . . inches*
	=	22.86	. . . . centimeters*

1 engineer's link	=	1	foot* (in tables)
1 Ramden's link	=	1	foot* (in tables)
1 cubit	=	18	inches*
	=	45.72	centimeters*
1 pace	=	2.5	feet*
	=	76.2	centimeters*
1 ell	=	45	inches*
	=	114.3	centimeters*
1 fathom	=	6	survey feet*
	=	1.828 804	meters
1 perch	=	1	rod* (in tables)
1 pole	=	1	rod* (in tables)
1 Gunter's chain	=	1	chain* (in tables)
1 surveyor's chain	=	1	chain* (in tables)
1 engineer's chain	=	100	feet*
	=	30.48	meters*
1 Ramden's chain	=	100	feet*
	=	30.48	meters*
1 bolt of cloth	=	120	feet*
	=	36.576	meters*
1 skein	=	360	feet*
	=	109.728	meters*
1 furlong	=	660	survey feet*
	=	201.168 4	meters
1 cable's length	=	720	survey feet*
	=	219.456 4	meters
1 international mile	=	1	mile* (in tables)
1 U.S. statute mile	=	1	survey mile* (in tables)
1 statute league	=	3	miles*
	=	4.828 032	kilometers*
1 land league	=	3	survey miles*
	=	4.828 042	kilometers

ASTRONOMICAL UNITS

1. The light year is the distance that light travels in 1 year. The speed of light in a vacuum equals 299 792 458 meters per second, according to Cohen and Taylor (ref 2). There are 31 536 000 seconds in a year of exactly 365 days. Thus:

$$1 \text{ light year} = 9.454\ 254\ 96 \times 10^{12} \text{ kilometers}$$
$$= 5.874\ 601\ 67 \times 10^{12} \text{ miles}$$

2. According to SI (ref 5), the astronomical unit of distance is the length of the radius of the unperturbed circular orbit of a body of negligible mass moving round the Sun with a sidereal angular velocity of 0.017 202 098 950 radian per day of 86 400 ephemeris seconds. In the system of astronomical constants of the International Astronomical Union the value adopted in 1976 is:

$$1 \text{ astronomical unit} = 149\ 597.870 \times 10^{6} \text{ meters}$$

3. Also according to SI (ref 5), one parsec is the distance at which 1 astronomical unit subtends an angle of 1 second of arc. Thus:

$$1 \text{ parsec} = 206\ 265 \qquad \text{astronomical units}$$
$$= \ \ 30\ 857 \times 10^{12} \text{ meters}$$

Light

LUMINANCE

One CANDELA PER SQUARE METER is equal to:

```
0.314 159 265 4 . . . . . . . . . . . . . millilambert
0.291 863 508 0 . . . . . . . . . . . . . footlambert
0.092 903 04 . . . . . . . candela per square foot*
0.000 645 16 . . . . . . . candela per square inch*
0.000 314 159 265 4 . . . . . . . . . . . . lambert

0.000 1 . . . . . . . . . . . . . . . . . . . . . . stilb*
```

One MILLILAMBERT is equal to:

```
3.183 098 862 . . . . . . candelas per square meter
0.929 030 4 . . . . . . . . . . . . . . footlambert*
0.295 719 560 9 . . . . . . . candela per square foot
0.002 053 608 062 . . . . . candela per square inch
0.001 . . . . . . . . . . . . . . . . . . . . . lambert*

0.000 318 309 886 2 . . . . . . . . . . . . . . stilb
```

One FOOTLAMBERT is equal to:

```
3.426 259 100 . . . . . . candelas per square meter
1.076 391 042 . . . . . . . . . . . . . millilamberts
0.318 309 886 2 . . . . . . . candela per square foot
0.002 210 485 321 . . . . . candela per square inch
0.001 076 391 042 . . . . . . . . . . . . . . lambert

0.000 342 625 910 0 . . . . . . . . . . . . . . stilb
```

One CANDELA PER SQUARE FOOT is equal to:

```
10.763 910 42 . . . . . . . candelas per square meter
```

3.381 582 189 . . . . millilamberts
3.141 592 654 . . . . footlamberts
0.006 944 444 444 . . . . . candela per square inch
0.003 381 582 189 . . . . lambert

0.001 076 391 042 . . . . stilb

One CANDELA PER SQUARE INCH is equal to:

1 550.003 100 . . . . candelas per square meter
486.947 835 2 . . . . millilamberts
452.389 342 1 . . . . footlamberts
144 . . . . candelas per square foot*
0.486 947 835 2 . . . . lambert

0.155 000 310 0 . . . . stilb

One LAMBERT is equal to:

3 183.098 862 . . . . candelas per square meter
1 000 . . . . millilamberts*
929.030 4 . . . . footlamberts*
295.719 560 9 . . . . candelas per square foot
2.053 608 062 . . . . candelas per square inch

0.318 309 886 2 . . . . stilb

One STILB or CANDELA PER SQUARE CENTIMETER is equal to:

10 000 . . . . candelas per square meter*
3 141.592 654 . . . . millilamberts
2 918.635 080 . . . . footlamberts
929.030 4 . . . . candelas per square foot*
6.451 6 . . . . candelas per square inch*

3.141 592 654 . . . . lamberts

ILLUMINANCE, ILLUMINATION

One LUX or LUMEN PER SQUARE METER is equal to:

0.1 . . . . milliphot*

0.092 903 04 . . . . footcandle*
0.000 1 . . . . phot*

One MILLIPHOT is equal to:

10 . . . . lux*
0.929 030 4 . . . . footcandle*
0.001 . . . . phot*

One FOOTCANDLE or LUMEN PER SQUARE FOOT is equal to:

10.763 910 42 . . . . lux
1.076 391 042 . . . . milliphots
0.001 076 391 042 . . . . phot

One PHOT or LUMEN PER SQUARE CENTIMETER is equal to:

10 000 . . . . lux*
1 000 . . . . milliphots*
929.030 4 . . . . footcandles*

ADDITIONAL CATEGORIES AND UNITS OF LIGHT

QUANTITY OF LIGHT   Basic unit: lumen-second

LIGHT EXPOSURE   Basic unit: lux-second

LUMINOUS EFFICACY   Basic unit: lumen per watt

LUMINOUS EXITANCE   Basic unit: lumen per square meter

Mass

One MILLIGRAM is equal to:

0.015 432 358 35 . . . . grain
0.001 . . . . gram*
0.000 771 617 917 6 . . . . . apothecaries scruple
0.000 643 014 931 4 . . . . pennyweight
0.000 564 383 391 2 . . . . avoirdupois dram

0.000 257 205 972 5 . . . . . . apothecaries dram
0.000 035 273 961 95 . . . . . avoirdupois ounce
0.000 032 150 746 57 . . . . apothecaries or troy
ounce
0.000 002 679 228 881 . . . . apothecaries or troy
pound
0.000 002 204 622 622 . . . . . avoirdupois pound

0.000 001 . . . . kilogram*
1.102 311 311 × 10⁻⁹ . . . . short ton
0.000 000 001 . . . . metric ton*
9.842 065 276 × 10⁻¹⁰ . . . . long ton

One GRAIN is equal to:

64.798 91 . . . . milligrams*
0.064 798 91 . . . . gram*
0.05 . . . . apothecaries scruple*
0.041 666 666 67 . . . . pennyweight
0.036 571 428 57 . . . . avoirdupois dram

0.016 666 666 67 . . . . apothecaries dram
0.002 285 714 286 . . . . avoirdupois ounce
0.002 083 333 333 . . . . apothecaries or troy
ounce
0.000 173 611 111 1 . . . . . apothecaries or troy
pound

133

0.000 142 857 142 9 . . . . . . avoirdupois pound

0.000 064 798 91 . . . . kilogram*
7.142 857 143 × 10⁻⁸ . . . . short ton
0.000 000 064 798 91 . . . . metric ton*
6.377 551 020 × 10⁻⁸ . . . . long ton

One GRAM is equal to:

1 000 . . . . milligrams*
15.432 358 35 . . . . grains
0.771 617 917 6 . . . . apothecaries scruple
0.643 014 931 4 . . . . pennyweight
0.564 383 391 2 . . . . avoirdupois dram

0.257 205 972 5 . . . . apothecaries dram
0.035 273 961 95 . . . . avoirdupois ounce
0.032 150 746 57 . . . . apothecaries or troy
ounce
0.002 679 228 881 . . . . apothecaries or troy
pound
0.002 204 622 622 . . . . avoirdupois pound

0.001 . . . . kilogram*
0.000 001 102 311 311 . . . . short ton
0.000 001 . . . . metric ton*
9.842 065 276 × 10⁻⁷ . . . . long ton

One APOTHECARIES SCRUPLE is equal to:

1 295.978 2 . . . . milligrams*
20 . . . . grains*
1.295 978 2 . . . . grams*
0.833 333 333 3 . . . . pennyweight
0.731 428 571 4 . . . . avoirdupois dram

0.333 333 333 3 . . . . apothecaries dram
0.045 714 285 71 . . . . avoirdupois ounce
0.041 666 666 67 . . . . apothecaries or troy
ounce
0.003 472 222 222 . . . . apothecaries or troy
pound

0.002 857 142 857 . . . . avoirdupois pound

0.001 295 978 2 . . . . kilogram*
0.000 001 428 571 429 . . . . short ton
0.000 001 295 978 2 . . . . metric ton*
0.000 001 275 510 204 . . . . long ton

One PENNYWEIGHT is equal to:

1 555.173 84 . . . . milligrams*
24 . . . . grains*
1.555 173 84 . . . . grams*
1.2 . . . . apothecaries scruples*
0.877 714 285 7 . . . . avoirdupois dram

0.4 . . . . apothecaries dram*
0.054 857 142 86 . . . . avoirdupois ounce
0.05 . . . . apothecaries or troy
                                                ounce*
0.004 166 666 667 . . . . apothecaries or troy
                                                pound
0.003 428 571 429 . . . . avoirdupois pound

0.001 555 173 84 . . . . kilogram*
0.000 001 714 285 714 . . . . short ton
0.000 001 555 173 84 . . . . metric ton*
0.000 001 530 612 245 . . . . long ton

One AVOIRDUPOIS DRAM is equal to:

1 771.845 195 312 5 . . . . milligrams*
27.343 75 . . . . grains*
1.771 845 195 312 5 . . . . grams*
1.367 187 5 . . . . apothecaries scruples*
1.139 322 917 . . . . pennyweights

0.455 729 166 7 . . . . apothecaries dram
0.062 5 . . . . avoirdupois ounce*
0.056 966 145 83 . . . . apothecaries or troy
                                                ounce
0.004 747 178 819 . . . . apothecaries or troy
                                                pound

```
0.003 906 25  . . . . . . . . . . . avoirdupois pound*
```

```
0.001 771 845 195 312 5  . . . . . . . kilogram*
0.000 001 953 125  . . . . . . . . . . . . short ton*
0.000 001 771 845 195 312 5  . . . metric ton*
0.000 001 743 861 607 . . . . . . . . . . . long ton
```

One **APOTHECARIES DRAM** is equal to:

```
3 887.934 6 . . . . . . . . . . . . . . . . . . . . milligrams*
    60  . . . . . . . . . . . . . . . . . . . . . . . . grains*
    3.887 934 6  . . . . . . . . . . . . . . . . . grams*
    3  . . . . . . . . . . . . . . . . apothecaries scruples*
    2.5  . . . . . . . . . . . . . . . . . . . . . pennyweights*
```

```
    2.194 285 714  . . . . . . . . . . avoirdupois drams
    0.137 142 857 1 . . . . . . . . . avoirdupois ounce
    0.125  . . . . . . . . . . . . . . . apothecaries or troy
                                                        ounce*
    0.010 416 666 67  . . . . . . . apothecaries or troy
                                                        pound
    0.008 571 428 571  . . . . . . . avoirdupois pound
```

```
    0.003 887 934 6 . . . . . . . . . . . . . . kilogram*
    0.000 004 285 714 286 . . . . . . . . . . short ton
    0.000 003 887 934 6  . . . . . . . . . metric ton*
    0.000 003 826 530 612 . . . . . . . . . . . long ton
```

One **AVOIRDUPOIS OUNCE** is equal to:

```
28 349.523 125  . . . . . . . . . . . . . . . . . . milligrams*
    437.5  . . . . . . . . . . . . . . . . . . . . . . . grains*
    28.349 523 125  . . . . . . . . . . . . . . . grams*
    21.875  . . . . . . . . . . . . . . apothecaries scruples*
    18.229 166 67  . . . . . . . . . . . . . . pennyweights
```

```
    16  . . . . . . . . . . . . . . . . . . . avoirdupois drams*
    7.291 666 667  . . . . . . . . . . apothecaries drams
    0.911 458 333 3 . . . . . . . . . apothecaries or troy
                                                        ounce
    0.075 954 861 11  . . . . . . . . apothecaries or troy
                                                        pound
```

```
0.062 5 . . . . . . . . . . . . . . . avoirdupois pound*

0.028 349 523 125 . . . . . . . . . . . . kilogram*
0.000 031 25 . . . . . . . . . . . . . . . short ton*
0.000 028 349 523 125 . . . . . . . . metric ton*
0.000 027 901 785 71 . . . . . . . . . . . long ton
```

One APOTHECARIES OR TROY OUNCE is equal to:

```
31 103.476 8 . . . . . . . . . . . . . . . . . . . . milligrams*
  480 . . . . . . . . . . . . . . . . . . . . . . . . . . . . grains*
   31.103 476 8 . . . . . . . . . . . . . . . . . . . grams*
   24 . . . . . . . . . . . . . . . . . apothecaries scruples*
   20 . . . . . . . . . . . . . . . . . . . . . . pennyweights*

   17.554 285 71 . . . . . . . . . . . . avoirdupois drams
    8 . . . . . . . . . . . . . . . . . . . . . apothecaries drams*
    1.097 142 857 . . . . . . . . . . avoirdupois ounces
    0.083 333 333 33 . . . . . . . . apothecaries or troy
                                                       pound
    0.068 571 428 57 . . . . . . . . avoirdupois pound

    0.031 103 476 8 . . . . . . . . . . . . . . kilogram*
    0.000 034 285 714 29 . . . . . . . . . . short ton
    0.000 031 103 476 8 . . . . . . . . . metric ton*
    0.000 030 612 244 90 . . . . . . . . . . . long ton
```

One APOTHECARIES OR TROY POUND is equal to:

```
373 241.721 6 . . . . . . . . . . . . . . . . . . . . milligrams*
  5 760 . . . . . . . . . . . . . . . . . . . . . . . . . . grains*
  373.241 721 6 . . . . . . . . . . . . . . . . . . grams*
  288 . . . . . . . . . . . . . . . . apothecaries scruples*
  240 . . . . . . . . . . . . . . . . . . . . . . pennyweights*

  210.651 428 6 . . . . . . . . . . . . avoirdupois drams
   96 . . . . . . . . . . . . . . . . . . apothecaries drams*
   13.165 714 29 . . . . . . . . . . avoirdupois ounces
   12 . . . . . . . . . . . . . . . . . apothecaries or troy
                                                      ounces*
    0.822 857 142 9 . . . . . . . . . . avoirdupois pound
```

```
0.373 241 721 6 . . . . . . . . . . . . . . kilogram*
0.000 411 428 571 4 . . . . . . . . . . . short ton
0.000 373 241 721 6 . . . . . . . . . metric ton*
0.000 367 346 938 8 . . . . . . . . . . . . long ton
```

One **AVOIRDUPOIS POUND** is equal to:

```
453 592.37 . . . . . . . . . . . . . . . . . . . milligrams*
    7 000 . . . . . . . . . . . . . . . . . . . . . . . grains*
  453.592 37 . . . . . . . . . . . . . . . . . . . . grams*
  350 . . . . . . . . . . . . . . . . apothecaries scruples*
  291.666 666 7 . . . . . . . . . . . . . . . pennyweights

  256 . . . . . . . . . . . . . . . . . . . avoirdupois drams*
  116.666 666 7 . . . . . . . . . . . apothecaries drams
   16 . . . . . . . . . . . . . . . . . . . . avoirdupois ounces*
   14.583 333 33 . . . . . . . . . . apothecaries or troy
                                                       ounces
    1.215 277 778 . . . . . . . . . apothecaries or troy
                                                       pounds

    0.453 592 37 . . . . . . . . . . . . . . . kilogram*
    0.000 5 . . . . . . . . . . . . . . . . . . . short ton*
    0.000 453 592 37 . . . . . . . . . . . . metric ton*
    0.000 446 428 571 4 . . . . . . . . . . . . long ton
```

One **KILOGRAM** is equal to:

```
1 000 000 . . . . . . . . . . . . . . . . . . . . . milligrams*
   15 432.358 35 . . . . . . . . . . . . . . . . . . . grains
    1 000 . . . . . . . . . . . . . . . . . . . . . . . grams*
  771.617 917 6 . . . . . . . . . . apothecaries scruples
  643.014 931 4 . . . . . . . . . . . . . . . pennyweights

  564.383 391 2 . . . . . . . . . . . . avoirdupois drams
  257.205 972 5 . . . . . . . . . . . apothecaries drams
   35.273 961 95 . . . . . . . . . . avoirdupois ounces
   32.150 746 57 . . . . . . . . . . apothecaries or troy
                                                       ounces
    2.679 228 881 . . . . . . . . . apothecaries or troy
                                                       pounds

    2.204 622 622 . . . . . . . . . . avoirdupois pounds
```

0.001 102 311 311 . . . . short ton
0.001 . . . . metric ton*
0.000 984 206 527 6 . . . . long ton

One SHORT TON is equal to:

907 184 740 . . . . milligrams*
14 000 000 . . . . grains*
907 184.74 . . . . grams*
700 000 . . . . apothecaries scruples*
583 333.333 3 . . . . pennyweights

512 000 . . . . avoirdupois drams*
233 333.333 3 . . . . apothecaries drams
32 000 . . . . avoirdupois ounces*
29 166.666 67 . . . . apothecaries or troy
ounces
2 430.555 556 . . . . apothecaries or troy
pounds

2 000 . . . . avoirdupois pounds*
907.184 74 . . . . kilograms*
0.907 184 74 . . . . metric ton*
0.892 857 142 9 . . . . long ton

One METRIC TON is equal to:

1 000 000 000 . . . . milligrams*
15 432 358.35 . . . . grains
1 000 000 . . . . grams*
771 617.917 6 . . . . apothecaries scruples
643 014.931 4 . . . . pennyweights

564 383.391 2 . . . . avoirdupois drams
257 205.972 5 . . . . apothecaries drams
35 273.961 95 . . . . avoirdupois ounces
32 150.746 57 . . . . apothecaries or troy
ounces
2 679.228 881 . . . . apothecaries or troy
pounds

2 204.622 622 . . . . avoirdupois pounds

```
1 000 . . . . . . . . . . . . . . . . . . . . . . . . kilograms*
      1.102 311 311 . . . . . . . . . . . . . . . short tons
      0.984 206 527 6 . . . . . . . . . . . . . . . . long ton
```

One LONG TON is equal to:

```
1 016 046 908.8 . . . . . . . . . . . . . . . . . . . . . milligrams*
   15 680 000 . . . . . . . . . . . . . . . . . . . . . . . . grains*
 1 016 046.908 8 . . . . . . . . . . . . . . . . . . . . . . grams*
      784 000 . . . . . . . . . . . . . . . . . apothecaries scruples*
      653 333.333 3 . . . . . . . . . . . . . . . . . . . pennyweights

      573 440 . . . . . . . . . . . . . . . . . . . avoirdupois drams*
      261 333.333 3 . . . . . . . . . . . . . . . apothecaries drams
       35 840 . . . . . . . . . . . . . . . . . . . avoirdupois ounces*
       32 666.666 67 . . . . . . . . . . . . . . . apothecaries or troy
                                                               ounces
    2 722.222 222 . . . . . . . . . . . . . . apothecaries or troy
                                                               pounds

    2 240 . . . . . . . . . . . . . . . . . . avoirdupois pounds*
 1 016.046 908 8 . . . . . . . . . . . . . . . . kilograms*
         1.12 . . . . . . . . . . . . . . . . . . . . . . . short tons*
      1.016 046 908 8 . . . . . . . . . . . . . metric tons*
```

ADDITIONAL UNITS OF MASS

Metric

1 gamma	=	1 . . . . microgram*
1 microgram	=	0.000 001 . . . . gram*
1 point	=	0.01 . . . . carat*
	=	2 . . . . milligrams*
1 centigram	=	0.01 . . . . gram*
1 decigram	=	0.1 . . . . gram*
1 carat	=	0.2 . . . . gram*
1 dekagram	=	10 . . . . grams*
1 assay ton	=	29.166 67 . . . . grams
1 hectogram	=	100 . . . . grams*
1 metric quintal	=	100 . . . . kilograms*

| 1 megagram | = | 1 . . . . metric ton* (in tables) |
| 1 tonne | = | 1 . . . . metric ton* (in tables) |

Customary

1 geepound or slug	=	32.174 05 . . . . avoirdupois pounds
	=	14.593 90 . . . . kilograms
1 cental	=	100 . . . . avoirdupois pounds*
	=	45.359 237 . . . . kilograms*
1 net or short hundred-weight	=	100 . . . . avoirdupois pounds*
	=	45.359 237 . . . . kilograms*
1 short quintal	=	100 . . . . avoirdupois pounds*
	=	45.359 237 . . . . kilograms*
1 gross or long hundred-weight	=	112 . . . . avoirdupois pounds*
	=	50.802 345 44 . . . . kilograms*
1 long quintal	=	112 . . . . avoirdupois pounds*
	=	50.802 345 44 . . . . kilograms*
1 short quarter	=	500 . . . . avoirdupois pounds*
	=	226.796 185 . . . . kilograms*
1 long quarter	=	560 . . . . avoirdupois pounds*
	=	254.011 727 2 . . . . kilograms*
1 net ton	=	1 . . . . short ton* (in tables)
1 gross ton	=	1 . . . . long ton* (in tables)
1 shipper's ton	=	1 . . . . long ton* (in tables)

Power

One ERG PER SECOND is equal to:

0.1 . . . . microwatt*

0.001 019 716 213 . . . . . gram-force centimeter per second

0.000 004 425 372 896 . . . foot pound-force per minute

6.118 297 278 $\times$ 10^{-7} . . . kilogram-force meter per minute

3.414 425 093 $\times$ 10^{-7} . . . . Btu per hour

0.000 000 1 . . . . watt*

7.375 621 493 $\times$ 10^{-8} . . . foot pound-force per second

2.390 057 361 $\times$ 10^{-8} . . . . . calorie per second

5.690 708 489 $\times$ 10^{-9} . . . . Btu per minute

1.434 034 417 $\times$ 10^{-9} . . kilocalorie per minute

1.359 621 617 $\times$ 10^{-10} . . . . . metric horsepower

1.341 022 090 $\times$ 10^{-10} . . . . horsepower

1.340 482 574 $\times$ 10^{-10} . . . . electric horsepower

0.000 000 000 1 . . . . kilowatt*

9.484 514 148 $\times$ 10^{-11} . . . . Btu per second

0.000 000 000 000 1 . . . . megawatt*

One MICROWATT is equal to:

10 . . . . ergs per second*

0.010 197 162 13 . . . . . . gram-force centimeter per second

0.000 044 253 728 96 . . . foot pound-force per minute

0.000 006 118 297 278 . . . kilogram-force meter per minute

142

0.000 003 414 425 093 . . . . Btu per hour

0.000 001 . . . . watt*
7.375 621 493 × 10^{-7} . . . foot pound-force per
second
2.390 057 361 × 10^{-7} . . . . . calorie per second
5.690 708 489 × 10^{-8} . . . . Btu per minute
1.434 034 417 × 10^{-8} . . kilocalorie per minute

1.359 621 617 × 10^{-9} . . . . . metric horsepower
1.341 022 090 × 10^{-9} . . . . horsepower
1.340 482 574 × 10^{-9} . . . . electric horsepower
0.000 000 001 . . . . kilowatt*
9.484 514 148 × 10^{-10} . . . . Btu per second

0.000 000 000 001 . . . . megawatt*

One GRAM-FORCE CENTIMETER PER SECOND is equal to:

980.665 . . . . ergs per second*
98.066 5 . . . . microwatts*
0.004 339 808 311 . . . . . . foot pound-force per
minute
0.000 6 . . . . kilogram-force meter per
minute*
0.000 334 840 718 4 . . . . Btu per hour

0.000 098 066 5 . . . . watt*
0.000 072 330 138 51 . . . foot pound-force per
second
0.000 023 438 456 02 . . . . . calorie per second
0.000 005 580 678 640 . . . . . . Btu per minute
0.000 001 406 307 361 . . kilocalorie per minute

1.333 333 333 × 10^{-7} . . . . . metric horsepower
1.315 093 427 × 10^{-7} . . . . horsepower
1.314 564 343 × 10^{-7} . . . . electric horsepower
0.000 000 098 066 5 . . . . kilowatt*
9.301 131 066 × 10^{-8} . . . . Btu per second

0.000 000 000 098 066 5 . . . . megawatt*

One FOOT POUND-FORCE PER MINUTE is equal to:

225 969.658 055 233 4 . . . . ergs per second*

22 596.965 805 523 34 . . . . microwatts*

230.424 923 96 . . . . . . gram-force centimeters per second*

0.138 254 954 376 . . . . kilogram-force meter per minute*

0.077 155 647 07 . . . . Btu per hour

0.022 596 965 805 523 34 . . . . watt*

0.016 666 666 67 . . . . foot pound-force per second

0.005 400 804 447 . . . . calorie per second

0.001 285 927 451 . . . . Btu per minute

0.000 324 048 266 8 . . . kilocalorie per minute

0.000 030 723 323 19 . . . . . metric horsepower

0.000 030 303 030 30 . . . . horsepower

0.000 030 290 838 88 . . . . electric horsepower

0.000 022 596 965 805 523 34 . . . . kilowatt*

0.000 021 432 124 19 . . . . Btu per second

0.000 000 022 596 965 805 523 34 . . . mega-watt*

One KILOGRAM-FORCE METER PER MINUTE is equal to:

1 634 441.667 . . . . ergs per second

163 444.166 7 . . . . microwatts

1 666.666 667 . . . . gram-force centimeters per second

7.233 013 851 . . . . foot pounds-force per minute

0.558 067 864 0 . . . . Btu per hour

0.163 444 166 7 . . . . watt

0.120 550 230 9 . . . . foot pound-force per second

0.039 064 093 37 . . . . calorie per second

0.009 301 131 066 . . . . Btu per minute

0.002 343 845 602 . . . . . kilocalorie per minute

```
0.000 222 222 222 2  . . . . . . metric horsepower
0.000 219 182 237 9  . . . . . . . . . . horsepower
0.000 219 094 057 2  . . . . . electric horsepower
0.000 163 444 166 7  . . . . . . . . . . . . kilowatt
0.000 155 018 851 1  . . . . . . . . Btu per second
```

$$1.634\ 441\ 667 \times 10^{-7} \quad . . . . \text{megawatt}$$

One **BTU PER HOUR** is equal to:

```
2 928 750.735  . . . . . . . . . . . . . . . . . . . ergs per second
  292 875.073 5 . . . . . . . . . . . . . . . . . . . . . microwatts
2 986.494 608 . . . . . . . . . gram-force centimeters per
                                                   second
  12.960 814 12  . . . . . . . . . foot pounds-force per
                                                    minute
   1.791 896 765  . . . . . . kilogram-force meters per
                                                    minute

   0.292 875 073 5 . . . . . . . . . . . . . . . . . . watt
   0.216 013 568 7 . . . . . . . . foot pound-force per
                                                    second
   0.069 998 822 53 . . . . . . . . calorie per second
   0.016 666 666 67 . . . . . . . . . . Btu per minute
   0.004 199 929 352 . . . . . kilocalorie per minute

   0.000 398 199 281 1  . . . . . . metric horsepower
   0.000 392 751 943 0  . . . . . . . . . horsepower
   0.000 392 593 932 3  . . . . . electric horsepower
   0.000 292 875 073 5  . . . . . . . . . . . . kilowatt
   0.000 277 777 777 8  . . . . . . . . Btu per second
```

$$2.928\ 750\ 735 \times 10^{-7} \quad . . . . \text{megawatt}$$

One **WATT** is equal to:

```
10 000 000  . . . . . . . . . . . . . . . . . . . . . . ergs per second*
 1 000 000  . . . . . . . . . . . . . . . . . . . . . . . microwatts*
10 197.162 13 . . . . . . . . . . gram-force centimeters per
                                                      second
   44.253 728 96  . . . . . . . . . . foot pounds-force per
                                                       minute
```

6.118 297 278 . . . . . . kilogram-force meters per
minute

3.414 425 093 . . . . Btu per hour
0.737 562 149 3 . . . . foot pound-force per
second
0.239 005 736 1 . . . . calorie per second
0.056 907 084 89 . . . . Btu per minute
0.014 340 344 17 . . . . . kilocalorie per minute

0.001 359 621 617 . . . . metric horsepower
0.001 341 022 090 . . . . horsepower
0.001 340 482 574 . . . . . . electric horsepower
0.001 . . . . kilowatt*
0.000 948 451 414 8 . . . . Btu per second

0.000 001 . . . . megawatt*

One FOOT POUND-FORCE PER SECOND is equal to:

13 558 179.483 314 004 . . . . ergs per second*
1 355 817.948 331 400 4 . . . . microwatts*
13 825.495 437 6 . . . . gram-force centimeters per
second*
60 . . . . foot pounds-force per minute*
8.295 297 262 56 . . . . kilogram-force meters per
minute*

4.629 338 824 . . . . Btu per hour
1.355 817 948 331 400 4 . . . . watts*
0.324 048 266 8 . . . . calorie per second
0.077 155 647 07 . . . . Btu per minute
0.019 442 896 01 . . . . . . kilocalorie per minute

0.001 843 399 391 68 . . . . metric horsepower*
0.001 818 181 818 . . . . horsepower
0.001 817 450 333 . . . . electric horsepower
0.001 355 817 948 331 400 4 . . . . . kilowatt*
0.001 285 927 451 . . . . Btu per second

0.000 001 355 817 948 331 400 4 . . . . megawatt*

One CALORIE PER SECOND is equal to:

41 840 000 . . . . ergs per second*
4 184 000 . . . . microwatts*
42 664.926 35 . . . . gram-force centimeters per
second
185.157 602 0 . . . . foot pounds-force per
minute
25.598 955 81 . . . . kilogram-force meters per
minute

14.285 954 59 . . . . Btu per hour
4.184 . . . . watts*
3.085 960 033 . . . . foot pounds-force per
second
0.238 099 243 2 . . . . Btu per minute
0.06 . . . . kilocalorie per minute*

0.005 688 656 847 . . . . metric horsepower
0.005 610 836 423 . . . . horsepower
0.005 608 579 088 . . . . electric horsepower
0.004 184 . . . . kilowatt*
0.003 968 320 719 . . . . Btu per second

0.000 004 184 . . . . megawatt*

One BTU PER MINUTE is equal to:

175 725 044.1 . . . . ergs per second
17 572 504.41 . . . . microwatts
179 189.676 5 . . . . gram-force centimeters per
second
777.648 847 2 . . . . foot pounds-force per
minute
107.513 805 9 . . . . kilogram-force meters per
minute

60 . . . . Btu per hour*
17.572 504 41 . . . . watts
12.960 814 12 . . . . foot pounds-force per
second
4.199 929 352 . . . . calories per second

0.251 995 761 1 . . . . kilocalorie per minute

0.023 891 956 86 . . . . metric horsepower
0.023 565 116 58 . . . . horsepower
0.023 555 635 94 . . . . electric horsepower
0.017 572 504 41 . . . . kilowatt
0.016 666 666 67 . . . . Btu per second

0.000 017 572 504 41 . . . . megawatt

One KILOCALORIE PER MINUTE is equal to:

697 333 333.3 . . . . ergs per second
69 733 333.33 . . . . microwatts
711 082.105 8 . . . . gram-force centimeters
per second
3 085.960 033 . . . . foot pounds-force per
minute
426.649 263 5 . . . . kilogram-force meters
per minute

238.099 243 2 . . . . Btu per hour
69.733 333 33 . . . . watts
51.432 667 21 . . . . foot pounds-force per
second
16.666 666 67 . . . . calories per second
3.968 320 719 . . . . Btu per minute

0.094 810 947 45 . . . . metric horsepower
0.093 513 940 38 . . . . horsepower
0.093 476 318 14 . . . . electric horsepower
0.069 733 333 33 . . . . kilowatt
0.066 138 678 66 . . . . Btu per second

0.000 069 733 333 33 . . . . megawatt

One METRIC HORSEPOWER is equal to:

7 354 987 500 . . . . ergs per second*
735 498 750 . . . . microwatts*
7 500 000 . . . . gram-force centimeters
per second*

32 548.562 33 . . . . foot pounds-force per
                                                    minute
4 500 . . . . kilogram-force meters
                                             per minute*

2 511.305 388 . . . . Btu per hour
735.498 75 . . . . watts*
542.476 038 8 . . . . foot pounds-force
                                                per second
175.788 420 2 . . . . calories per second
41.855 089 80 . . . . Btu per minute

10.547 305 21 . . . . kilocalories per minute
0.986 320 070 6 . . . . horsepower
0.985 923 257 4 . . . . electric horsepower
0.735 498 75 . . . . kilowatt*
0.697 584 830 0 . . . . Btu per second

0.000 735 498 75 . . . . megawatt*

One HORSEPOWER is equal to:

7 456 998 715.822 702 2 . . . . ergs per second*
745 699 871.582 270 22 . . . . microwatts*
7 604 022.490 68 . . . . gram-force centimeters
                                                    per second*
33 000 . . . . foot pounds-force per
                                                       minute*
4 562.413 494 408 . . . . kilogram-force meters
                                                per minute*

2 546.136 353 . . . . Btu per hour
745.699 871 582 270 22 . . . . watts*
550 . . . . foot pounds-force per
                                                     second*
178.226 546 7 . . . . calories per second
42.435 605 89 . . . . Btu per minute

10.693 592 80 . . . . kilocalories per minute
1.013 869 665 424 . . . . metric horsepower*
0.999 597 683 1 . . . . electric horsepower
0.745 699 871 582 270 22 . . . . kilowatt*

0.707 260 098 2 . . . . Btu per second

0.000 745 699 871 582 270 22 . . . . megawatt*

One ELECTRIC HORSEPOWER is equal to:

7 460 000 000 . . . . ergs per second*

746 000 000 . . . . microwatts*

7 607 082.949 . . . . gram-force centimeters
per second

33 013.281 80 . . . . foot pounds-force per
minute

4 564.249 769 . . . . kilogram-force meters per
minute

2 547.161 119 . . . . Btu per hour

746 . . . . watts*

550.221 363 4 . . . . foot pounds-force
per second

178.298 279 2 . . . . calories per second

42.452 685 32 . . . . Btu per minute

10.697 896 75 . . . . kilocalories per minute

1.014 277 727 . . . . metric horsepower

1.000 402 479 . . . . horsepower

0.746 . . . . kilowatt*

0.707 544 755 4 . . . . Btu per second

0.000 746 . . . . megawatt*

One KILOWATT is equal to:

10 000 000 000 . . . . ergs per second*

1 000 000 000 . . . . microwatts*

10 197 162.13 . . . . gram-force centimeters
per second

44 253.728 96 . . . . foot pounds-force per
minute

6 118.297 278 . . . . kilogram-force meters
per minute

3 414.425 093 . . . . Btu per hour

1 000 . . . . watts*

737.562 149 3 . . . . foot pounds-force per
second

239.005 736 1 . . . . calories per second

56.907 084 89 . . . . Btu per minute

14.340 344 17 . . . . kilocalories per minute

1.359 621 617 . . . . metric horsepower

1.341 022 090 . . . . horsepower

1.340 482 574 . . . . electric horsepower

0.948 451 414 8 . . . . Btu per second

0.001 . . . . megawatt*

One BTU PER SECOND is equal to:

1.054 350 264 × 10¹⁰ . . . . ergs per second

1.054 350 264 × 10⁹ . . . . microwatts

10 751 380.59 . . . . gram-force centimeters
per second

46 658.930 83 . . . . foot pounds-force per
minute

6 450.828 353 . . . . kilogram-force meters
per minute

3 600 . . . . Btu per hour*

1 054.350 264 . . . . watts

777.648 847 2 . . . . foot pounds-force per
second

251.995 761 1 . . . . calories per second

60 . . . . Btu per minute*

15.119 745 67 . . . . kilocalories per minute

1.433 517 412 . . . . metric horsepower

1.413 906 995 . . . . horsepower

1.413 338 156 . . . . electric horsepower

1.054 350 264 . . . . kilowatts

0.001 054 350 264 . . . . megawatt

One MEGAWATT is equal to:

10 000 000 000 000 . . . . ergs per second*

1 000 000 000 000 .... microwatts*

1.019 716 213 × 10^{10} .. gram-force centimeters per second

44 253 728.96 .... foot pounds-force per minute

6 118 297.278 .... kilogram-force meters per minute

3 414 425.093 .... Btu per hour

1 000 000 .... watts*

737 562.149 3 .... foot pounds-force per second

239 005.736 1 .... calories per second

56 907.084 89 .... Btu per minute

14 340.344 17 .... kilocalories per minute

1 359.621 617 .... metric horsepower

1 341.022 090 .... horsepower

1 340.482 574 .... electric horsepower

1 000 .... kilowatts*

948.451 414 8 .... Btu per second

ADDITIONAL UNITS OF POWER

(Abbreviations used are I.T. for International Steam Table and thermo. for thermochemical.)

Metric

1 gram-force centi- meter per hour	=	2.724 069 × 10^{-8} .... watt
1 gram-force centi- meter per minute	=	1.634 442 × 10^{-6} .... watt
1 thermochemical gram calorie/hour	=	0.001 162 222 .... watt
1 I.T. gram calorie per hour	=	0.001 163 .... watt*
1 mean gram calorie per hour	=	0.001 163 89 .... watt

1 kilogram-force meter
per hour = 0.002 724 069 . . . . watt

1 thermo. gram calorie
per minute = 0.069 733 33 . . . . watt
1 I.T. gram calorie
per minute = 0.069 78 . . . . watt*
1 mean gram calorie
per minute = 0.069 833 7 . . . . watt

1 thermo. kilogram
calorie/hour = 1.162 222 . . . . watts
1 I.T. kilogram
calorie/hour = 1.163 . . . . watts*
1 mean kilogram
calorie/hour = 1.163 89 . . . . watts

1 thermo. gram
calorie per second = 1 . . . . calorie/second*
(in tables)
= 4.184 . . . . watts*
1 I.T. gram calorie
per second = 4.186 8 . . . . watts*
1 mean gram calorie
per second = 4.190 02 . . . . watts

1 kilogram-force meter
per second = 9.806 65 . . . . watts*

1 thermo. kilogram
calorie/minute = 1 . . . . kilocalorie per minute*
(in tables)
= 69.733 33 . . . . watts
1 I.T. kilogram
calorie/minute = 69.78 . . . . watts*
1 mean kilogram
calorie/minute = 69.833 7 . . . . watts

1 hectowatt = 100 . . . . watts*
1 cheval-vapeur = 1 . . . . metric horsepower*
(in tables)
= 735.498 75 . . . . watts*

1 thermo. kilogram
calorie/second = 4 184 . . . . watts*

1 I.T. kilogram
  calorie/second   = 4 186.8 . . . . watts*
1 mean kilogram
  calorie/second   = 4 190.02 . . . . watts

Customary

1 thermochemical
  Btu per hour   =   1 . . . . Btu per hour*
                                                                (in tables)
                        =   0.292 875 1 . . . . watt
1 I.T. Btu per hour   =   0.293 071 1 . . . . watt
1 mean Btu per hour   =   0.293 296 . . . . watt

1 thermochemical Btu
  per minute   =   1 . . . . Btu per minute*
                                                                (in tables)
                        =   17.572 50 . . . . watts
1 I.T. Btu per minute   =   17.584 26 . . . . watts
1 mean Btu per minute   =   17.597 8 . . . . watts

1 thermochemical Btu
  per second   =   1 . . . . Btu per second*
                                                                (in tables)
                        = 1 054.350 . . . . watts
1 I.T. Btu per second   = 1 055.056 . . . . watts
1 mean Btu per second   = 1 055.87 . . . . watts
1 ton of refrigeration =   200 . . . . I.T. Btu per minute*
                        = 3 516.853 . . . . watts

1 boiler horsepower   =   34.5 . . . . pound of water per
                                                     hour evaporated from
                                                     and at 212°F
                        = 9 809.50 . . . . watts

Pressure

One DYNE PER SQUARE CENTIMETER is equal to:

0.1 . . . . pascal*
0.010 197 162 13 . . . . kilogram-force per
square meter
0.002 088 543 423 . . . . pound-force per
square foot
0.001 019 716 213 . . . . . . gram-force per square
centimeter
0.001 . . . . millibar*

0.000 750 062 7 . . . . millimeter of mercury
0.000 401 474 3 . . . . inch of water
0.000 1 . . . . kilopascal*
0.000 033 456 19 . . . . foot of water
0.000 029 530 03 . . . . inch of mercury

0.000 014 503 773 77 . . . . . . pound-force per
square inch
0.000 010 197 162 13 . . . . metric ton-force per
square meter
0.000 01 . . . . newton per square centimeter*
0.000 001 044 271 712 . . . . . short ton-force per
square foot
0.000 001 019 716 213 . . . kilogram-force/square
centimeter

0.000 001 . . . . bar*
9.869 232 667 × 10^{-7} . . . . atmosphere

One PASCAL is equal to:

1 . . . . newton per square meter*

155

10 . . . . dynes per square centimeter*
0.101 971 621 3 . . . . kilogram-force per
square meter
0.020 885 434 23 . . . . . . pound-force per square
foot
0.010 197 162 13 . . . . gram-force per square
centimeter

0.01 . . . . millibar*
0.007 500 627 . . . . millimeter of mercury
0.004 014 743 . . . . inch of water
0.001 . . . . kilopascal*
0.000 334 561 9 . . . . foot of water

0.000 295 300 3 . . . . inch of mercury
0.000 145 037 737 7 . . . . pound-force per
square inch
0.000 101 971 621 3 . . . . . metric ton-force per
square meter
0.000 1 . . . . newton per square centimeter*
0.000 010 442 717 12 . . . . . short ton-force per
square foot

0.000 010 197 162 13 . . . kilogram-force/square
centimeter
0.000 01 . . . . bar*
0.000 009 869 232 667 . . . . atmosphere

One **KILOGRAM-FORCE PER SQUARE METER** is equal to:

98.066 5 . . . . dynes per square centimeter*
9.806 65 . . . . pascals*
0.204 816 143 6 . . . . pound-force per square
foot
0.1 . . . . gram-force per square centimeter*
0.098 066 5 . . . . millibar*
0.073 556 02 . . . . millimeter of mercury

0.039 371 18 . . . . inch of water
0.009 806 65 . . . . kilopascal*
0.003 280 932 . . . . foot of water
0.002 895 906 . . . . inch of mercury

0.001 422 334 331 . . . . . pound-force per square
inch
0.001 . . . . metric ton-force per square
meter*
0.000 980 665 . . . . newton per square centimeter*
0.000 102 408 071 8 . . . . . . short ton-force per
square foot
0.000 1 . . . . kilogram-force per square
centimeter*

0.000 098 066 5 . . . . bar*
0.000 096 784 110 54 . . . . atmosphere

One **POUND-FORCE PER SQUARE FOOT** is equal to:

478.802 589 8 . . . . dynes per square centimeter
47.880 258 98 . . . . pascals
4.882 427 636 . . . . . . kilograms-force per square
meter
0.488 242 763 6 . . . . gram-force per square
centimeter
0.478 802 589 8 . . . . millibar

0.359 132 0 . . . . millimeter of mercury
0.192 226 94 . . . . inch of water
0.047 880 258 98 . . . . kilopascal
0.016 018 912 . . . . foot of water
0.014 139 053 . . . . inch of mercury

0.006 944 444 444 . . . . . pound-force per square
inch
0.004 882 427 636 . . metric ton-force per square
meter
0.004 788 025 898 . . . . newton per square centimeter

0.000 5 . . . . short ton-force per square
foot*

0.000 488 242 763 6 . . . . kilogram-force/square
centimeter

0.000 478 802 589 8 . . . . bar

0.000 472 541 416 0 . . . . atmosphere

One GRAM-FORCE PER SQUARE CENTIMETER is equal to:

980.665 . . . . dynes per square centimeter*

98.066 5 . . . . pascals*

10 . . . . kilograms-force per square
meter*

2.048 161 436 . . . . pounds-force per square
foot

0.980 665 . . . . millibar*

0.735 560 2 . . . . millimeter of mercury

0.393 711 8 . . . . inch of water

0.098 066 5 . . . . kilopascal*

0.032 809 32 . . . . foot of water

0.028 959 06 . . . . inch of mercury

0.014 223 343 31 . . . . . . pound-force per square
inch

0.01 . . . . metric ton-force per square
meter*

0.009 806 65 . . . . newton per square centimeter*

0.001 024 080 718 . . . short ton-force per square
foot

0.001 . . . . kilogram-force per square centimeter*

0.000 980 665 . . . . bar*

0.000 967 841 105 4 . . . . atmosphere

One MILLIBAR is equal to:

1 000 . . . . dynes per square centimeter*

100 . . . . pascals*
10.197 162 13 . . . . . . kilograms-force per square
meter
2.088 543 423 . . . . pounds-force per square
foot
1.019 716 213 . . . . grams-force per square
centimeter

0.750 062 7 . . . . millimeter of mercury
0.401 474 3 . . . . inch of water
0.1 . . . . kilopascal*
0.033 456 19 . . . . foot of water
0.029 530 03 . . . . inch of mercury

0.014 503 773 77 . . . . . . pound-force per square
inch
0.010 197 162 13 . . . metric ton-force per square
meter
0.01 . . . . newton per square centimeter*
0.001 044 271 712 . . . short ton-force per square
foot
0.001 019 716 213 . . . kilogram-force per square
centimeter

0.001 . . . . bar*
0.000 986 923 266 7 . . . . atmosphere

One **MILLIMETER OF MERCURY** (at 0°C) is equal to:

1 333.221 9 . . . . dynes per square centimeter
133.322 19 . . . . pascals
13.595 080 . . . . kilograms-force per square
meter
2.784 492 . . . . pounds-force per square
foot
1.359 508 0 . . . . grams-force per square
centimeter

1.333 221 9 . . . . millibars
0.535 254 4 . . . . inch of water
0.133 322 19 . . . . kilopascal
0.044 604 53 . . . . foot of water

0.039 370 078 74 . . . . inch of mercury

0.019 336 749 . . . . pound-force per square
inch

0.013 595 080 . . . . . metric ton-force per square
meter

0.013 332 219 . . . . newton per square centimeter

0.001 392 245 9 . . . . short ton-force per
square foot

0.001 359 508 0 . . . . . kilogram-force per square
centimeter

0.001 333 221 9 . . . . bar

0.001 315 787 7 . . . . atmosphere

One INCH OF WATER (at 3.98°C, its maximum density) is equal to:

2 490.819 . . . . dynes per square centimeter

249.081 9 . . . . pascals

25.399 29 . . . . kilograms-force per square
meter

5.202 184 . . . . pounds-force per square
foot

2.539 929 . . . . grams-force per square
centimeter

2.490 819 . . . . millibars

1.868 270 6 . . . . millimeters of mercury

0.249 081 9 . . . . kilopascal

0.083 333 333 33 . . . . foot of water

0.073 553 96 . . . . inch of mercury

0.036 126 28 . . . . pound-force per square
inch

0.025 399 29 . . . . metric ton-force per
square meter

0.024 908 19 . . . . newton per square centimeter

0.002 601 092 . . . . . . short ton-force per square
foot

0.002 539 929 . . . . . . kilogram-force per square
                                     centimeter

0.002 490 819 . . . . bar
0.002 458 248 . . . . atmosphere

One KILOPASCAL is equal to:

10 000 . . . . dynes per square centi-
                                     meter*
1 000 . . . . pascals*
101.971 621 3 . . . . kilograms-force per square
                                     meter
20.885 434 23 . . . . pounds-force per square
                                     foot
10.197 162 13 . . . . grams-force per square
                                     centimeter

10 . . . . millibars*
7.500 627 . . . . millimeters of mercury
4.014 743 . . . . inches of water
0.334 561 9 . . . . foot of water
0.295 300 3 . . . . inch of mercury

0.145 037 737 7 . . . . pound-force per square
                                     inch
0.101 971 621 3 . . . . metric ton-force per
                                     square meter
0.1 . . . . newton per square centi-
                                     meter*
0.010 442 717 12 . . . . short ton-force per
                                     square foot
0.010 197 162 13 . . . . kilogram-force per square
                                     centimeter

0.01 . . . . bar*
0.009 869 232 667 . . . . atmosphere

One FOOT OF WATER (at 3.98°C, its maximum density) is equal to:

29 889.83 . . . . dynes per square centi-
                                     meter

2 988.983 . . . . pascals

304.791 5 . . . . kilograms-force per square
meter

62.426 21 . . . . pounds-force per square
foot

30.479 15 . . . . grams-force per square
centimeter

29.889 83 . . . . millibars

22.419 25 . . . . millimeters of mercury

12 . . . . inches of water*

2.988 983 . . . . kilopascals

0.882 647 5 . . . . inch of mercury

0.433 515 4 . . . . pound-force per square
inch

0.304 791 5 . . . . metric ton-force per square
meter

0.298 898 3 . . . . newton per square centimeter

0.031 213 11 . . . . short ton-force per square
foot

0.030 479 15 . . . . kilogram-force per square
centimeter

0.029 889 83 . . . . bar

0.029 498 97 . . . . atmosphere

One INCH OF MERCURY (at 0°C) is equal to:

33 863.84 . . . . dynes per square centimeter

3 386.384 . . . . pascals

345.315 0 . . . . kilograms-force per square
meter

70.726 09 . . . . pounds-force per square
foot

34.531 50 . . . . grams-force per square
centimeter

33.863 84 . . . . millibars

25.4 . . . . millimeters of mercury*

13.595 461 . . . . inches of water
3.386 384 . . . . kilopascals
1.132 955 1 . . . . feet of water

0.491 153 4 . . . . pound-force per square
inch
0.345 315 0 . . . . metric ton-force per square
meter
0.338 638 4 . . . . newton per square centimeter
0.035 363 05 . . . . short ton-force per square
foot
0.034 531 50 . . . . kilogram-force per square
centimeter

0.033 863 84 . . . . bar
0.033 421 01 . . . . atmosphere

One **POUND-FORCE PER SQUARE INCH** is equal to:

68 947.572 93 . . . . dynes per square centimeter
6 894.757 293 . . . . pascals
703.069 579 6 . . . . kilograms-force per square
meter
144 . . . . pounds-force per square
foot*
70.306 957 96 . . . . grams-force per square
centimeter

68.947 572 93 . . . . millibars
51.715 00 . . . . millimeters of mercury
27.680 68 . . . . inches of water
6.894 757 293 . . . . kilopascals
2.306 723 . . . . feet of water

2.036 024 . . . . inches of mercury
0.703 069 579 6 . . . . metric ton-force per square
meter
0.689 475 729 3 . . . . . . newton per square centimeter

0.072 . . . . short ton-force per square
foot*

0.070 306 957 96 . . . . kilogram-force per square
centimeter

0.068 947 572 93 . . . . bar
0.068 045 963 91 . . . . atmosphere

One METRIC TON-FORCE PER SQUARE METER is equal to:

98 066.5 . . . . dynes per square centimeter*

9 806.65 . . . . pascals*

1 000 . . . . kilograms-force per square
meter*

204.816 143 6 . . . . pounds-force per square
foot

100 . . . . grams-force per square centimeter*

98.066 5 . . . . millibars*
73.556 02 . . . . millimeters of mercury
39.371 18 . . . . inches of water
9.806 65 . . . . kilopascals*
3.280 932 . . . . feet of water

2.895 906 . . . . inches of mercury
1.422 334 331 . . . . pounds-force per square
inch

0.980 665 . . . . newton per square centimeter*

0.102 408 071 8 . . . . . short ton-force per square
foot

0.1 . . . . kilogram-force per square centimeter*

0.098 066 5 . . . . bar*
0.096 784 110 54 . . . . atmosphere

One NEWTON PER SQUARE CENTIMETER is equal to:

100 000 . . . . dynes per square centimeter*

10 000 .... pascals*
1 019.716 213 .... kilograms-force per square
meter
208.854 342 3 .... pounds-force per square
foot
101.971 621 3 .... grams-force per square
centimeter

100 .... millibars*
75.006 27 .... millimeters of mercury
40.147 43 .... inches of water
10 .... kilopascals*
3.345 619 .... feet of water

2.953 003 .... inches of mercury
1.450 377 377 .... pounds-force per square
inch
1.019 716 213 .... metric tons-force per square
meter
0.104 427 171 2 ..... short ton-force per square
foot
0.101 971 621 3 ..... kilogram-force per square
centimeter

0.1 .... bar*
0.098 692 326 67 .... atmosphere

One **SHORT TON-FORCE PER SQUARE FOOT** is equal to:

957 605.179 6 .... dynes per square centimeter
95 760.517 96 .... pascals
9 764.855 273 .... kilograms-force per square
meter
2 000 .... pounds-force per square
foot*
976.485 527 3 .... grams-force per square
centimeter

957.605 179 6 .... millibars
718.263 9 .... millimeters of mercury
384.453 9 .... inches of water
95.760 517 96 .... kilopascals

32.037 82 . . . . feet of water

28.278 11 . . . . inches of mercury
13.888 888 89 . . . . pounds-force per square
inch
9.764 855 273 . . . . metric tons-force per square
meter
9.576 051 796 . . . . . . newtons per square centimeter
0.976 485 527 3 . . . . . kilogram-force per square
centimeter

0.957 605 179 6 . . . . bar
0.945 082 832 1 . . . . atmosphere

One **KILOGRAM-FORCE PER SQUARE CENTIMETER** is equal to:

980 665 . . . . dynes per square centimeter*
98 066.5 . . . . pascals*
10 000 . . . . kilograms-force per square
meter*
2 048.161 436 . . . . pounds-force per square
foot
1 000 . . . . grams-force per square
centimeter*

980.665 . . . . millibars*
735.560 2 . . . . millimeters of mercury
393.711 8 . . . . inches of water
98.066 5 . . . . kilopascals*
32.809 32 . . . . feet of water

28.959 06 . . . . inches of mercury
14.223 343 31 . . . . pounds-force per square
inch
10 . . . . metric tons-force per square
meter*
9.806 65 . . . . newtons per square centimeter*
1.024 080 718 . . . . . short tons-force per square
foot

0.980 665 . . . . bar*
0.967 841 105 4 . . . . atmosphere

One BAR is equal to:

1 000 000 . . . . dynes per square centimeter*
100 000 . . . . pascals*
10 197.162 13 . . . . kilograms-force per square
meter
2 088.543 423 . . . . pounds-force per square
foot
1 019.716 213 . . . . grams-force per square
centimeter

1 000 . . . . millibars*
750.062 7 . . . . millimeters of mercury
401.474 3 . . . . inches of water
100 . . . . kilopascals*
33.456 19 . . . . feet of water

29.530 03 . . . . inches of mercury
14.503 773 77 . . . . pounds-force per square
inch
10.197 162 13 . . . . . metric tons-force per square
meter
10 . . . . newtons per square centimeter*
1.044 271 712 . . . . . short tons-force per square
foot

1.019 716 213 . . . . . . kilograms-force per square
centimeter
0.986 923 266 7 . . . . atmosphere

One ATMOSPHERE is equal to:

1 013 250 . . . . dynes per square centimeter*
101 325 . . . . pascals*
10 332.274 53 . . . . kilograms-force per square
meter

2 116.216 624 . . . . pounds-force per square
foot
1 033.227 453 . . . . grams-force per square
centimeter

1 013.25 . . . . millibars*
760.001 0 . . . . millimeters of mercury
406.793 9 . . . . inches of water
101.325 . . . . kilopascals*
33.899 49 . . . . feet of water

29.921 30 . . . . inches of mercury
14.695 948 78 . . . . pounds-force per square
inch
10.332 274 53 . . . . metric tons-force per
square meter
10.132 5 . . . . newtons per square centimeter*

1.058 108 312 . . . . . short tons-force per square
foot

1.033 227 453 . . . . . . kilograms-force per square
centimeter
1.013 25 . . . . bars*

ADDITIONAL UNITS OF PRESSURE

Metric

1 dyne per square meter	=	0.000 01 . . . . pascal*
1 gram-force per square meter	=	0.009 806 65 . . . . pascal*
1 barye	=	1 . . . . dyne per square centi-meter* (in tables)
	=	0.1 . . . . pascal*
1 newton per square meter	=	1 . . . . pascal* (in tables)
1 dyne per square millimeter	=	10 . . . . pascals*

1 gram-force per square millimeter	=	1 . . . . metric ton-force/sq meter* (in tables)
	=	9 806.65 . . . . pascals*
1 decibar	=	1 . . . . newton per square centimeter* (in tables)
	=	10 000 . . . . pascals*
1 technical atmosphere	=	1 . . . . kilogram-force/sq centimeter* (in tables)
	=	98 066.5 . . . . pascals*
1 normal atmosphere	=	1 . . . . atmosphere* (in tables)
	=	101 325 . . . . pascals*
1 newton per square millimeter	=	1 000 000 . . . . pascals*

Customary

1 poundal per square foot	=	1.488 164 . . . . pascals
1 ounce-force per square foot	=	2.992 516 . . . . pascals
1 poundal per square inch	=	214.295 6 . . . . pascals
1 ounce-force per square inch	=	430.922 3 . . . . pascals
1 psi	=	1 . . . . pound-force per square inch* (in tables)
	=	6 894.757 . . . . pascals
1 kip per square foot	=	47 880.26 . . . . pascals
1 long ton-force per square foot	=	107 251.8 . . . . pascals
1 kip per square inch	=	$6.894\ 757 \times 10^6$ . . . . pascals
1 short ton-force per square inch	=	$1.378\ 951 \times 10^7$ . . . . pascals
1 long ton-force per square inch	=	$1.544\ 426 \times 10^7$ . . . . pascals

Other Units

1 centimeter of water at 3.98°C	=	98.063 75 . . . . pascals
1 torr	=	1 . . . . millimeter of mercury at 3.98°C* (in tables)
	=	133.322 19 . . . . pascals
1 inch of water at 60°F	=	248.84 . . . . pascals
1 centimeter of mercury at 0°C	=	1 333.221 9 . . . . pascals
1 inch of mercury at 60°F	=	3 376.85 . . . . pascals
1 foot of mercury at 0°C	=	40 636.60 . . . . pascals
1 meter of mercury at 0°C	=	133 322.19 . . . . pascals

Temperature

The temperature scales in general use are the Celsius, in degrees C or °C, and the Fahrenheit, in degrees F or °F. The Celsius scale was formerly called centigrade.

As established by SI, the kelvin is the unit of thermodynamic temperature and has the symbol K instead of "degree Kelvin" (symbol °K). The triple point of water is assigned the temperature 273.16 K by definition. Also, 0°C is defined as 273.15 K.

The Rankine scale, in degrees R or °R, is an absolute scale used for scientific purposes and is related to the Fahrenheit scale.

Absolute zero temperature has the following exact values on the four scales:

$$\text{absolute zero temperature} = 0 \text{ K} = 0°\text{R} = -273.15°\text{C} = -459.67°\text{F}$$

The following equations are used to convert temperatures, and are presented with the correct symbols and subscripts:

T_K = temperature in kelvins $\quad t°_C$ = temperature in °C

$t°_R$ = temperature in °R $\quad t°_F$ = temperature in °F

$$t°_C = \frac{5}{9}(t°_F - 32) \quad = \frac{5}{9}(t°_R - 491.67) \quad = T_K - 273.15$$

$$t°_F = \frac{9}{5}t°_C + 32 \quad = \frac{9}{5}T_K - 459.67 \quad = t°_R - 459.67$$

$$T_K = t°_C + 273.15 \quad = \frac{5}{9}(t°_F + 459.67) \quad = \frac{5}{9}t°_R$$

$$t°_R = t°_F + 459.67 \quad = \frac{9}{5}t°_C + 491.67 \quad = \frac{9}{5}T_K$$

The following five tables are designed to simplify conversions between the Fahrenheit and Celsius scales. Some numbers are given to five decimal places to nullify rounding errors when adding values, although the measurement of temperature is generally limited to only two or three decimal places. Examples are presented to illustrate the use of the tables.

Table 1. Exact Equivalents Between the Fahrenheit and Celsius Scales

$t°_F$	$t°_C$	$t°_F$	$t°_C$	$t°_F$	$t°_C$	$t°_F$	$t°_C$	$t°_F$	$t°_C$	$t°_F$	$t°_C$	$t°_F$	$t°_C$
-454	-270	-229	-145	-4	-20	221	105	446	230	671	355	896	480
-445	-265	-220	-140	5	-15	230	110	455	235	680	360	905	485
-436	-260	-211	-135	14	-10	239	115	464	240	689	365	914	490
-427	-255	-202	-130	23	-5	248	120	473	245	698	370	923	495
-418	-250	-193	-125	32	0	257	125	482	250	707	375	932	500
-409	-245	-184	-120	41	5	266	130	491	255	716	380	941	505
-400	-240	-175	-115	50	10	275	135	500	260	725	385	950	510
-391	-235	-166	-110	59	15	284	140	509	265	734	390	959	515
-382	-230	-157	-105	68	20	293	145	518	270	743	395	968	520
-373	-225	-148	-100	77	25	302	150	527	275	752	400	977	525
-364	-220	-139	-95	86	30	311	155	536	280	761	405	986	530
-355	-215	-130	-90	95	35	320	160	545	285	770	410	995	535
-346	-210	-121	-85	104	40	329	165	554	290	779	415	1004	540
-337	-205	-112	-80	113	45	338	170	563	295	788	420	1013	545
-328	-200	-103	-75	122	50	347	175	572	300	797	425	1022	550
-319	-195	-94	-70	131	55	356	180	581	305	806	430	1031	555
-310	-190	-85	-65	140	60	365	185	590	310	815	435	1040	560
-301	-185	-76	-60	149	65	374	190	599	315	824	440	1049	565
-292	-180	-67	-55	158	70	383	195	608	320	833	445	1058	570
-283	-175	-58	-50	167	75	392	200	617	325	842	450	1067	575
-274	-170	-49	-45	176	80	401	205	626	330	851	455	1076	580
-265	-165	-40	-40	185	85	410	210	635	335	860	460	1085	585
-256	-160	-31	-35	194	90	419	215	644	340	869	465	1094	590
-247	-155	-22	-30	203	95	428	220	653	345	878	470	1103	595
-238	-150	-13	-25	212	100	437	225	662	350	887	475	1112	600

Table 2. Converting Intervals of 1 to 9°F to °C

°F	°C	°F	°C
1	0.55556	6	3.33333
2	1.11111	7	3.88889
3	1.66667	8	4.44444
4	2.22222	9	5*
5	2.77778		

Table 3. Converting Intervals of 1 to 5°C to °F

°C	°F
1	1.8*
2	3.6*
3	5.4*
4	7.2*
5	9 *

Table 4. Converting Intervals of Hundredths of 1°F to °C

°F	°C	°F	°C	°F	°C	°F	°C	°F	°C
.0	.0*	.2	.11111	.4	.22222	.6	.33333	.8	.44444
.01	.00556	.21	.11667	.41	.22778	.61	.33889	.81	.45*
.02	.01111	.22	.12222	.42	.23333	.62	.34444	.82	.45556
.03	.01667	.23	.12778	.43	.23889	.63	.35*	.83	.46111
.04	.02222	.24	.13333	.44	.24444	.64	.35556	.84	.46667
.05	.02778	.25	.13889	.45	.25*	.65	.36111	.85	.47222
.06	.03333	.26	.14444	.46	.25556	.66	.36667	.86	.47778
.07	.03889	.27	.15*	.47	.26111	.67	.37222	.87	.48333
.08	.04444	.28	.15556	.48	.26667	.68	.37778	.88	.48889
.09	.05*	.29	.16111	.49	.27222	.69	.38333	.89	.49444
.1	.05556	.3	.16667	.5	.27778	.7	.38889	.9	.5*
.11	.06111	.31	.17222	.51	.28333	.71	.39444	.91	.50556
.12	.06667	.32	.17778	.52	.28889	.72	.4*	.92	.51111
.13	.07222	.33	.18333	.53	.29444	.73	.40556	.93	.51667
.14	.07778	.34	.18889	.54	.3*	.74	.41111	.94	.52222
.15	.08333	.35	.19444	.55	.30556	.75	.41667	.95	.52778
.16	.08889	.36	.2*	.56	.31111	.76	.42222	.96	.53333
.17	.09444	.37	.20556	.57	.31667	.77	.42778	.97	.53889
.18	.1*	.38	.21111	.58	.32222	.78	.43333	.98	.54444
.19	.10556	.39	.21667	.59	.32778	.79	.43889	.99	.55*

All relationships are exact in Table 5

Example 1. Convert the normal human body temperature of 98.6°F to $t_{°C}$.

In Table 1 find $t_{°F}$ closest to and less than 98.6 with an exact equivalent in $t_{°C}$: 95°F = 35°C

In Table 2 find the interval in °C equal to the difference between 95 and 98°F, which is 3°F: 1.66667°C

In Table 4 find the interval in °C equal to 0.6°F: 0.33333°C

Add the three values in °C: 35

$$\begin{array}{r} 1.66667 \\ \underline{0.33333} \\ 37.00000 \end{array}$$

Round off the answer to read 37.0°C.

Table 5. Converting Intervals of Hundredths of 1°C to °F

$°C$	$°F$	$°C$	$°F$	$°C$	$°F$	$°C$	$°F$	$°C$	$°F$
.0	.0	.2	.36	.4	.72	.6	1.08	.8	1.44
.01	.018	.21	.378	.41	.738	.61	1.098	.81	1.458
.02	.036	.22	.396	.42	.756	.62	1.116	.82	1.476
.03	.054	.23	.414	.43	.774	.63	1.134	.83	1.494
.04	.072	.24	.432	.44	.792	.64	1.152	.84	1.512
.05	.09	.25	.45	.45	.81	.65	1.17	.85	1.53
.06	.108	.26	.468	.46	.828	.66	1.188	.86	1.548
.07	.126	.27	.486	.47	.846	.67	1.206	.87	1.566
.08	.144	.28	.504	.48	.864	.68	1.224	.88	1.584
.09	.162	.29	.522	.49	.882	.69	1.242	.89	1.602
.1	.18	.3	.54	.5	.9	.7	1.26	.9	1.62
.11	.198	.31	.558	.51	.918	.71	1.278	.91	1.638
.12	.216	.32	.576	.52	.936	.72	1.296	.92	1.656
.13	.234	.33	.594	.53	.954	.73	1.314	.93	1.674
.14	.252	.34	.612	.54	.972	.74	1.332	.94	1.692
.15	.27	.35	.63	.55	.99	.75	1.35	.95	1.71
.16	.288	.36	.648	.56	1.008	.76	1.368	.96	1.728
.17	.306	.37	.666	.57	1.026	.77	1.386	.97	1.746
.18	.324	.38	.684	.58	1.044	.78	1.404	.98	1.764
.19	.342	.39	.702	.59	1.062	.79	1.422	.99	1.782

Example 2.   Convert absolute zero temperature, $-273.15°C$, to $t°_F$.

In Table 1 find $t°_C$ closest to and greater than -273 with an exact equivalent in $t°_F$: $-270°C = -454°F$

In Table 3 find the interval in °F equal to the difference between −273 and −270°C, which is 3°C: 5.4°F

In Table 5 find the interval in °F equal to 0.15°C: 0.27°F

Add the three values in °F: −454

−   5.4

−   0.27

−459.67

Thus absolute zero is equal to −459.67°F, exactly.

Time

One SECOND is equal to:

$$0.016\ 666\ 666\ 67\ \dots\dots\dots\dots\dots\text{minute}$$
$$0.000\ 277\ 777\ 777\ 8\ \dots\dots\dots\dots\dots\text{hour}$$
$$0.000\ 011\ 574\ 074\ 07\ \dots\dots\dots\dots\dots\text{day}$$
$$0.000\ 001\ 653\ 439\ 153\ \dots\dots\dots\dots\dots\text{week}$$
$$4.133\ 597\ 884\times10^{-7}\ \dots\dots\dots\text{28-day month}$$

$$3.991\ 060\ 026\times10^{-7}\ \dots\dots\dots\text{29-day month}$$
$$3.858\ 024\ 691\times10^{-7}\ \dots\dots\dots\text{30-day month}$$
$$3.733\ 572\ 282\times10^{-7}\ \dots\dots\dots\text{31-day month}$$
$$3.170\ 979\ 198\times10^{-8}\ \dots\dots\dots\dots\dots\text{year}$$
$$3.162\ 315\ 321\times10^{-8}\ \dots\dots\dots\dots\text{leap year}$$

One MINUTE is equal to:

$$60\ \dots\dots\dots\dots\dots\dots\dots\text{seconds*}$$
$$0.016\ 666\ 666\ 67\ \dots\dots\dots\dots\dots\text{hour}$$
$$0.000\ 694\ 444\ 444\ 4\ \dots\dots\dots\dots\dots\text{day}$$
$$0.000\ 099\ 206\ 349\ 21\ \dots\dots\dots\dots\dots\text{week}$$
$$0.000\ 024\ 801\ 587\ 30\ \dots\dots\dots\text{28-day month}$$

$$0.000\ 023\ 946\ 360\ 15\ \dots\dots\dots\text{29-day month}$$
$$0.000\ 023\ 148\ 148\ 15\ \dots\dots\dots\text{30-day month}$$
$$0.000\ 022\ 401\ 433\ 69\ \dots\dots\dots\text{31-day month}$$
$$0.000\ 001\ 902\ 587\ 519\ \dots\dots\dots\dots\dots\text{year}$$
$$0.000\ 001\ 897\ 389\ 192\ \dots\dots\dots\dots\text{leap year}$$

One HOUR is equal to:

$$3\ 600\ \dots\dots\dots\dots\dots\dots\dots\text{seconds*}$$
$$60\ \dots\dots\dots\dots\dots\dots\dots\text{minutes*}$$
$$0.041\ 666\ 666\ 67\ \dots\dots\dots\dots\dots\text{day}$$
$$0.005\ 952\ 380\ 952\ \dots\dots\dots\dots\dots\text{week}$$

0.001 488 095 238 . . . . 28-day month

0.001 436 781 609 . . . . 29-day month
0.001 388 888 889 . . . . 30-day month
0.001 344 086 022 . . . . 31-day month
0.000 114 155 251 1 . . . . year
0.000 113 843 351 5 . . . . leap year

One DAY is equal to:

86 400 . . . . seconds*
1 440 . . . . minutes*
24 . . . . hours*
0.142 857 142 9 . . . . week
0.035 714 285 71 . . . . 28-day month

0.034 482 758 62 . . . . 29-day month
0.033 333 333 33 . . . . 30-day month
0.032 258 064 52 . . . . 31-day month
0.002 739 726 027 . . . . year
0.002 732 240 437 . . . . leap year

One WEEK is equal to:

604 800 . . . . seconds*
10 080 . . . . minutes*
168 . . . . hours*
7 . . . . days*
0.25 . . . . 28-day month*

0.241 379 310 3 . . . . 29-day month
0.233 333 333 3 . . . . 30-day month
0.225 806 451 6 . . . . 31-day month
0.019 178 082 19 . . . . year
0.019 125 683 06 . . . . leap year

One 28-DAY MONTH is equal to:

2 419 200 . . . . seconds*
40 320 . . . . minutes*
672 . . . . hours*

28 . . . . days*
4 . . . . weeks*

0.965 517 241 4 . . . . 29-day month
0.933 333 333 3 . . . . 30-day month
0.903 225 806 5 . . . . 31-day month
0.076 712 328 77 . . . . year
0.076 502 732 24 . . . . leap year

One 29-DAY MONTH is equal to:

2 505 600 . . . . seconds*
41 760 . . . . minutes*
696 . . . . hours*
29 . . . . days*
4.142 857 143 . . . . weeks

1.035 714 286 . . . . 28-day months
0.966 666 666 7 . . . . 30-day month
0.935 483 871 0 . . . . 31-day month
0.079 452 054 79 . . . . year
0.079 234 972 68 . . . . leap year

One 30-DAY MONTH is equal to:

2 592 000 . . . . seconds*
43 200 . . . . minutes*
720 . . . . hours*
30 . . . . days*
4.285 714 286 . . . . weeks

1.071 428 571 . . . . 28-day months
1.034 482 759 . . . . 29-day months
0.967 741 935 5 . . . . 31-day month
0.082 191 780 82 . . . . year
0.081 967 213 11 . . . . leap year

One 31-DAY MONTH is equal to:

2 678 400 . . . . seconds*
44 640 . . . . minutes*

```
744  . . . . . . . . . . . . . . . . . . . . . . . . . . . hours*
 31  . . . . . . . . . . . . . . . . . . . . . . . . . . . days*
4.428 571 429  . . . . . . . . . . . . . . . . . . weeks

1.107 142 857  . . . . . . . . . . . . 28-day months
1.068 965 517  . . . . . . . . . . . . 29-day months
1.033 333 333  . . . . . . . . . . . . 30-day months
0.084 931 506 85 . . . . . . . . . . . . . . . . year
0.084 699 453 55 . . . . . . . . . . . . . leap year
```

One YEAR is equal to:

```
31 536 000  . . . . . . . . . . . . . . . . . . . . . . . . . . seconds*
   525 600  . . . . . . . . . . . . . . . . . . . . . . . . . . minutes*
     8 760  . . . . . . . . . . . . . . . . . . . . . . . . . . . hours*
       365  . . . . . . . . . . . . . . . . . . . . . . . . . . . days*
52.142 857 14  . . . . . . . . . . . . . . . . . . . weeks

13.035 714 29  . . . . . . . . . . . . . 28-day months
12.586 206 90  . . . . . . . . . . . . . 29-day months
12.166 666 67  . . . . . . . . . . . . . 30-day months
11.774 193 55  . . . . . . . . . . . . . 31-day months
0.997 267 759 6 . . . . . . . . . . . . . . . leap year
```

One LEAP YEAR is equal to:

```
31 622 400  . . . . . . . . . . . . . . . . . . . . . . . . . . seconds*
   527 040  . . . . . . . . . . . . . . . . . . . . . . . . . . minutes*
     8 784  . . . . . . . . . . . . . . . . . . . . . . . . . . . hours*
       366  . . . . . . . . . . . . . . . . . . . . . . . . . . . days*
52.285 714 29  . . . . . . . . . . . . . . . . . . . weeks

13.071 428 57  . . . . . . . . . . . . . 28-day months
12.620 689 66  . . . . . . . . . . . . . 29-day months
12.2  . . . . . . . . . . . . . . . . . . 30-day months*
11.806 451 61  . . . . . . . . . . . . . 31-day months
1.002 739 726  . . . . . . . . . . . . . . . . . . years
```

ADDITIONAL UNITS OF TIME

```
1 shake        = 0.000 000 01  . . . . . . . . . . . . . . . . . . . second*
```

One solar day is the time of one revolution of the earth on its axis as measured by the interval between two successive transits of the Sun over the same meridian. Because solar days are of unequal duration the mean solar day represents the average.

The sidereal day is the time of one revolution of the earth on its axis relative to the position of the earth with respect to the stars.

The solar or tropical year is the time for the earth to complete one revolution around the Sun based on two consecutive returns of the Sun to the vernal equinox.

The sidereal year is the time for the earth to complete one revolution around the Sun based on the position of the earth with respect to a given star, that is, the time for the earth to return to the same position with respect to the same star.

Velocity

One FOOT PER MINUTE is equal to:

0.508 . . . . centimeter per second*
0.304 8 . . . . meter per minute*
0.018 288 . . . . kilometer per hour*
0.016 666 666 67 . . . . foot per second
0.011 363 636 36 . . . . mile per hour

0.009 874 730 022 . . . . knot
0.005 08 . . . . meter per second*
0.000 304 8 . . . . kilometer per minute*
0.000 189 393 939 4 . . . . mile per minute
0.000 003 156 565 657 . . . . mile per second

One CENTIMETER PER SECOND is equal to:

1.968 503 937 . . . . feet per minute
0.6 . . . . meter per minute*
0.036 . . . . kilometer per hour*
0.032 808 398 95 . . . . foot per second
0.022 369 362 92 . . . . mile per hour

0.019 438 444 92 . . . . knot
0.01 . . . . meter per second*
0.000 6 . . . . kilometer per minute*
0.000 372 822 715 3 . . . . mile per minute
0.000 006 213 711 922 . . . . mile per second

One METER PER MINUTE is equal to:

3.280 839 895 . . . . feet per minute
1.666 666 667 . . . . centimeters per second
0.06 . . . . kilometer per hour*
0.054 680 664 92 . . . . foot per second

181

0.037 282 271 53 . . . . mile per hour

0.032 397 408 21 . . . . knot
0.016 666 666 67 . . . . meter per second
0.001 . . . . kilometer per minute*
0.000 621 371 192 2 . . . . mile per minute
0.000 010 356 186 54 . . . . mile per second

One KILOMETER PER HOUR is equal to:

54.680 664 92 . . . . feet per minute
27.777 777 78 . . . . centimeters per second
16.666 666 67 . . . . meters per minute
0.911 344 415 3 . . . . foot per second
0.621 371 192 2 . . . . mile per hour

0.539 956 803 5 . . . . knot
0.277 777 777 8 . . . . meter per second
0.016 666 666 67 . . . . . . kilometer per minute
0.010 356 186 54 . . . . mile per minute
0.000 172 603 109 0 . . . . mile per second

One FOOT PER SECOND is equal to:

60 . . . . feet per minute*
30.48 . . . . centimeters per second*
18.288 . . . . meters per minute*
1.097 28 . . . . kilometers per hour*
0.681 818 181 8 . . . . mile per hour

0.592 483 801 3 . . . . knot
0.304 8 . . . . meter per second*
0.018 288 . . . . kilometer per minute*
0.011 363 636 36 . . . . mile per minute
0.000 189 393 939 4 . . . . mile per second

One MILE PER HOUR is equal to:

88 . . . . feet per minute*
44.704 . . . . centimeters per second*
26.822 4 . . . . meters per minute*

1.609 344 . . . . kilometers per hour*
1.466 666 667 . . . . feet per second

0.868 976 241 9 . . . . knot
0.447 04 . . . . meter per second*
0.026 822 4 . . . . kilometer per minute*
0.016 666 666 67 . . . . mile per minute
0.000 277 777 777 8 . . . . mile per second

One KNOT is equal to:

101.268 591 4 . . . . feet per minute
51.444 444 44 . . . . centimeters per second
30.866 666 67 . . . . meters per minute
1.852 . . . . kilometers per hour*
1.687 809 857 . . . . feet per second

1.150 779 448 . . . . miles per hour
0.514 444 444 4 . . . . meter per second
0.030 866 666 67 . . . . . . kilometer per minute
0.019 179 657 47 . . . . mile per minute
0.000 319 660 957 8 . . . . mile per second

One METER PER SECOND is equal to:

196.850 393 7 . . . . feet per minute
100 . . . . centimeters per second*
60 . . . . meters per minute*
3.6 . . . . kilometers per hour*
3.280 839 895 . . . . feet per second

2.236 936 292 . . . . miles per hour
1.943 844 492 . . . . knots
0.06 . . . . kilometer per minute*
0.037 282 271 53 . . . . mile per minute
0.000 621 371 192 2 . . . . mile per second

One KILOMETER PER MINUTE is equal to:

3 280.839 895 . . . . feet per minute
1 666.666 667 . . . . centimeters per second

1 000 . . . . meters per minute*
60 . . . . kilometers per hour*
54.680 664 92 . . . . feet per second

37.282 271 53 . . . . miles per hour
32.397 408 21 . . . . knots
16.666 666 67 . . . . meters per second
0.621 371 192 2 . . . . mile per minute
0.010 356 186 54 . . . . mile per second

One MILE PER MINUTE is equal to:

5 280 . . . . feet per minute*
2 682.24 . . . . centimeters per second*
1 609.344 . . . . meters per minute*
96.560 64 . . . . kilometers per hour*
88 . . . . feet per second*

60 . . . . miles per hour*
52.138 574 51 . . . . knots
26.822 4 . . . . meters per second*
1.609 344 . . . . kilometers per minute*
0.016 666 666 67 . . . . mile per second

One MILE PER SECOND is equal to:

316 800 . . . . feet per minute*
160 934.4 . . . . centimeters per second*
96 560.64 . . . . meters per minute*
5 793.638 4 . . . . kilometers per hour*
5 280 . . . . feet per second*

3 600 . . . . miles per hour*
3 128.314 471 . . . . knots
1 609.344 . . . . meters per second*
96.560 64 . . . . kilometers per minute*
60 . . . . miles per minute*

ADDITIONAL UNITS OF VELOCITY

Metric

1 centimeter per minute	=	0.000 166 666 7 . . . . . . meter per second
	=	0.000 6 . . . . kilometer per hour*
1 meter per hour	=	0.000 277 777 8 . . . . . . meter per second
	=	0.001 . . . . kilometer per hour*
1 international knot	=	1 . . . . knot* (in tables)
	=	0.514 444 4 . . . . meter per second
	=	1.852 . . . . kilometers per hour*
1 kilometer per second	=	1 000 . . . . meters per second*
	=	3 600 . . . . kilometers per hour*

Customary

1 inch per hour	=	0.000 007 055 556 . . . . . . meter/second
	=	0.000 025 4 . . . . kilometer per hour*
1 foot per hour	=	0.000 084 666 67 . . . . meter/second
	=	0.000 304 8 . . . . kilometer per hour*
1 yard per hour	=	0.000 254 . . . . meter per second*
	=	0.000 914 4 . . . . kilometer per hour*
1 inch per minute	=	0.000 423 333 3 . . . . . . meter per second
	=	0.001 524 . . . . kilometer per hour*
1 yard per minute	=	0.015 24 . . . . meter per second*
	=	0.054 864 . . . . kilometer per hour*
1 inch per second	=	0.025 4 . . . . meter per second*
	=	0.091 44 . . . . kilometer per hour*
1 U.S. statute mile per hour	=	1 . . . . survey mile per hour*
	=	0.447 040 9 . . . . meter per second
	=	1.609 347 . . . . kilometers per hour
1 yard per second	=	0.914 4 . . . . meter per second*
	=	3.291 84 . . . . kilometers per hour*
1 U.S. statute mile per minute	=	1 . . . . survey mile per minute*
	=	26.822 45 . . . . meters per second
	=	96.560 83 . . . . kilometers per hour

1 U.S. statute mile
    per second   =   1 . . . . survey mile per second*
                      = 1 609.347 . . . . meters per second
                      = 5 793.650 . . . . kilometers per hour

Volume

One MINIM is equal to:

0.061 611 519 921 875 . . . . cubic centimeter or
milliliter*
0.016 666 666 67 . . . . fluid dram
0.003 759 765 625 . . . . cubic inch*
0.002 083 333 333 . . . . fluid ounce
0.000 520 833 333 3 . . . . gill

0.000 130 208 333 3 . . . . liquid pint
0.000 111 896 745 8 . . . . dry pint
0.000 065 104 166 67 . . . . liquid quart
0.000 061 611 519 921 875 . . . cubic decimeter
or liter*
0.000 055 948 372 88 . . . . dry quart

0.000 016 276 041 67 . . . . gallon
0.000 006 993 546 609 . . . . peck
0.000 002 175 790 292 . . . . cubic foot
0.000 001 748 386 652 . . . . bushel
3.875 248 016 X 10^{-7} . . . . . . petroleum barrel

8.058 482 564 X 10^{-8} . . . . cubic yard
0.000 000 061 611 519 921 875 . . cubic meter
or kiloliter*

One CUBIC CENTIMETER or MILLILITER is equal to:

16.230 730 90 . . . . minims
0.270 512 181 6 . . . . fluid dram
0.061 023 744 09 . . . . cubic inch
0.033 814 022 70 . . . . fluid ounce
0.008 453 505 675 . . . . gill

0.002 113 376 419 . . . . liquid pint

187

```
0.001 816 165 969 . . . . . . . . . . . . . . dry pint
0.001 056 688 209 . . . . . . . . . . . liquid quart
0.001 . . . . . . . . . . . . . cubic decimeter or liter*
0.000 908 082 984 3 . . . . . . . . . . . dry quart

0.000 264 172 052 4 . . . . . . . . . . . . . gallon
0.000 113 510 373 0 . . . . . . . . . . . . . . peck
0.000 035 314 666 72 . . . . . . . . . cubic foot
0.000 028 377 593 26 . . . . . . . . . . . bushel
0.000 006 289 810 770 . . . . . petroleum barrel

0.000 001 307 950 619 . . . . . . . . . cubic yard
0.000 001 . . . . . . . . . . cubic meter or kiloliter*
```

One FLUID DRAM is equal to:

```
60 . . . . . . . . . . . . . . . . . . . . . . . . . minims*
3.696 691 195 312 5 . . . . . cubic centimeters or
                                        milliliters*
0.225 585 937 5 . . . . . . . . . . . . . cubic inch*
0.125 . . . . . . . . . . . . . . . . . . . . fluid ounce*
0.031 25 . . . . . . . . . . . . . . . . . . . . . . gill*

0.007 812 5 . . . . . . . . . . . . . . . . liquid pint*
0.006 713 804 745 . . . . . . . . . . . . . . dry pint
0.003 906 25 . . . . . . . . . . . . . . liquid quart*
0.003 696 691 195 312 5 . . . . . cubic decimeter
                                        or liter*
0.003 356 902 373 . . . . . . . . . . . . . dry quart

0.000 976 562 5 . . . . . . . . . . . . . . . . gallon*
0.000 419 612 796 6 . . . . . . . . . . . . . . peck
0.000 130 547 417 5 . . . . . . . . . . cubic foot
0.000 104 903 199 1 . . . . . . . . . . . . bushel
0.000 023 251 488 10 . . . . . . petroleum barrel

0.000 004 835 089 538 . . . . . . . . . . cubic yard
0.000 003 696 691 195 312 5 . . . cubic meter
                                        or kiloliter*
```

One CUBIC INCH is equal to:

```
265.974 026 0 . . . . . . . . . . . . . . . . . . . minims
```

16.387 064 . . . . . . cubic centimeters or milliliters*
4.432 900 433 . . . . fluid drams
0.554 112 554 1 . . . . fluid ounce
0.138 528 138 5 . . . . gill

0.034 632 034 63 . . . . liquid pint
0.029 761 627 96 . . . . dry pint
0.017 316 017 32 . . . . liquid quart
0.016 387 064 . . . . cubic decimeter or liter*
0.014 880 813 98 . . . . dry quart

0.004 329 004 329 . . . . gallon
0.001 860 101 748 . . . . peck
0.000 578 703 703 7 . . . . cubic foot
0.000 465 025 436 9 . . . . bushel
0.000 103 071 531 6 . . . . petroleum barrel

0.000 021 433 470 51 . . . . cubic yard
0.000 016 387 064 . . . . cubic meter or kiloliter*

One FLUID OUNCE is equal to:

480 . . . . minims*
29.573 529 562 5 . . . . . cubic centimeters or milli-
                                                    liters*
8 . . . . fluid drams*
1.804 687 5 . . . . cubic inches*
0.25 . . . . gill*

0.062 5 . . . . liquid pint*
0.053 710 437 96 . . . . dry pint
0.031 25 . . . . liquid quart*
0.029 573 529 562 5 . . . . . . cubic decimeter or
                                                    liter*
0.026 855 218 98 . . . . dry quart

0.007 812 5 . . . . gallon*
0.003 356 902 373 . . . . peck
0.001 044 379 340 . . . . cubic foot
0.000 839 225 593 1 . . . . bushel
0.000 186 011 904 8 . . . . petroleum barrel

0.000 038 680 716 31 . . . . cubic yard
0.000 029 573 529 562 5 . . . . . cubic meter or
kiloliter*

One GILL is equal to:

1 920 . . . . minims*
118.294 118 25 . . . . cubic centimeters or milliliters*
32 . . . . fluid drams*
7.218 75 . . . . cubic inches*
4 . . . . fluid ounces*

0.25 . . . . liquid pint*
0.214 841 751 8 . . . . dry pint
0.125 . . . . liquid quart*
0.118 294 118 25 . . . . . cubic decimeter or liter*
0.107 420 875 9 . . . . dry quart

0.031 25 . . . . gallon*
0.013 427 609 49 . . . . peck
0.004 177 517 361 . . . . cubic foot
0.003 356 902 373 . . . . bushel
0.000 744 047 619 0 . . . . petroleum barrel

0.000 154 722 865 2 . . . . cubic yard
0.000 118 294 118 25 . . . . cubic meter or kiloliter*

One LIQUID PINT is equal to:

7 680 . . . . minims*
473.176 473 . . . . cubic centimeters or milliliters*
128 . . . . fluid drams*
28.875 . . . . cubic inches*
16 . . . . fluid ounces*

4 . . . . gills*
0.859 367 007 4 . . . . dry pint
0.5 . . . . liquid quart*

0.473 176 473 . . . . cubic decimeter or liter*
0.429 683 503 7 . . . . dry quart

0.125 . . . . gallon*
0.053 710 437 96 . . . . peck
0.016 710 069 44 . . . . cubic foot
0.013 427 609 49 . . . . bushel
0.002 976 190 476 . . . . petroleum barrel

0.000 618 891 460 9 . . . . cubic yard
0.000 473 176 473 . . . . cubic meter or kiloliter*

One DRY PINT is equal to:

8 936.810 390 . . . . minims
550.610 471 357 5 . . . . . cubic centimeters or milliliters*
148.946 839 8 . . . . fluid drams
33.600 312 5 . . . . cubic inches*
18.618 354 98 . . . . fluid ounces

4.654 588 745 . . . . gills
1.163 647 186 . . . . liquid pints
0.581 823 593 1 . . . . liquid quart
0.550 610 471 357 5 . . cubic decimeter or liter*
0.5 . . . . dry quart*

0.145 455 898 3 . . . . gallon
0.062 5 . . . . peck*
0.019 444 625 29 . . . . cubic foot
0.015 625 . . . . bushel*
0.003 463 235 673 . . . . petroleum barrel

0.000 720 171 307 0 . . . . cubic yard
0.000 550 610 471 357 5 . . cubic meter or kiloliter*

One LIQUID QUART is equal to:

15 360 . . . . minims*
946.352 946 . . . . . . cubic centimeters or milliliters*
256 . . . . fluid drams*
57.75 . . . . cubic inches*

32 . . . . fluid ounces*

8 . . . . gills*
2 . . . . liquid pints*
1.718 734 015 . . . . dry pints
0.946 352 946 . . . . cubic decimeter or liter*
0.859 367 007 4 . . . . dry quart

0.25 . . . . gallon*
0.107 420 875 9 . . . . peck
0.033 420 138 89 . . . . cubic foot
0.026 855 218 98 . . . . bushel
0.005 952 380 952 . . . . petroleum barrel

0.001 237 782 922 . . . . cubic yard
0.000 946 352 946 . . . . cubic meter or kiloliter*

One CUBIC DECIMETER or LITER is equal to:

16 230.730 90 . . . . minims
1 000 . . . . cubic centimeters or milliliters*
270.512 181 6 . . . . fluid drams
61.023 744 09 . . . . cubic inches
33.814 022 70 . . . . fluid ounces

8.453 505 675 . . . . gills
2.113 376 419 . . . . liquid pints
1.816 165 969 . . . . dry pints
1.056 688 209 . . . . liquid quarts
0.908 082 984 3 . . . . dry quart

0.264 172 052 4 . . . . gallon
0.113 510 373 0 . . . . peck
0.035 314 666 72 . . . . cubic foot
0.028 377 593 26 . . . . bushel
0.006 289 810 770 . . . . petroleum barrel

0.001 307 950 619 . . . . cubic yard
0.001 . . . . cubic meter or kiloliter*

One DRY QUART is equal to:

17 873.620 78 . . . . minims

1 101.220 942 715 . . . . . . cubic centimeters or milliliters*
297.893 679 7 . . . . fluid drams
67.200 625 . . . . cubic inches*
37.236 709 96 . . . . fluid ounces

9.309 177 489 . . . . gills
2.327 294 372 . . . . liquid pints
2 . . . . dry pints*
1.163 647 186 . . . . liquid quarts
1.101 220 942 715 . . . . cubic decimeters or
liters*

0.290 911 796 5 . . . . gallon
0.125 . . . . peck*
0.038 889 250 58 . . . . cubic foot
0.031 25 . . . . bushel*
0.006 926 471 346 . . . . petroleum barrel

0.001 440 342 614 . . . . cubic yard
0.001 101 220 942 715 . . . . cubic meter or kiloliter*

One GALLON is equal to:

61 440 . . . . minims*
3 785.411 784 . . . . cubic centimeters or milliliters*
1 024 . . . . fluid drams*
231 . . . . cubic inches*
128 . . . . fluid ounces*

32 . . . . gills*
8 . . . . liquid pints*
6.874 936 059 . . . . dry pints
4 . . . . liquid quarts*
3.785 411 784 . . . . . . cubic decimeters or liters*

3.437 468 030 . . . . dry quarts
0.429 683 503 7 . . . . peck
0.133 680 555 6 . . . . cubic foot
0.107 420 875 9 . . . . bushel
0.023 809 523 81 . . . . petroleum barrel

0.004 951 131 687 . . . . cubic yard
0.003 785 411 784 . . . . cubic meter or kiloliter*

One PECK is equal to:

142 988.966 2. . . . . mimims
8 809.767 541 72 . . . . cubic centimeters or milliliters*
2 383.149 437 . . . . fluid drams
537.605 . . . . cubic inches*
297.893 679 7 . . . . fluid ounces

74.473 419 91 . . . . gills
18.618 354 98 . . . . liquid pints
16 . . . . dry pints*
9.309 177 489 . . . . liquid quarts
8.809 767 541 72 . . . . cubic decimeters or liters*

8 . . . . dry quarts*
2.327 294 372 . . . . gallons
0.311 114 004 6 . . . . cubic foot
0.25 . . . . bushel*
0.055 411 770 77 . . . . petroleum barrel

0.011 522 740 91 . . . . cubic yard
0.008 809 767 541 72 . . . . cubic meter or kiloliter*

One CUBIC FOOT is equal to:

459 603.116 9. . . . . minims
28 316.846 592 . . . . cubic centimeters or milliliters*
7 660.051 948 . . . . fluid drams
1 728 . . . . cubic inches*
957.506 493 5 . . . . fluid ounces

239.376 623 4 . . . . gills
59.844 155 84 . . . . liquid pints
51.428 093 12 . . . . dry pints
29.922 077 92 . . . . liquid quarts

28.316 846 592 . . . . . . cubic decimeters or liters*

25.714 046 56 . . . . dry quarts
7.480 519 481 . . . . gallons
3.214 255 820 . . . . pecks
0.803 563 954 9 . . . . bushel
0.178 107 606 7 . . . . petroleum barrel

0.037 037 037 04 . . . . cubic yard
0.028 316 846 592 . . . . cubic meter or kiloliter*

One BUSHEL is equal to:

571 955.864 9 . . . . minims
35 239.070 166 88 . . . . cubic centimeters or milliliters*
9 532.597 749 . . . . fluid drams
2 150.42 . . . . cubic inches*
1 191.574 719 . . . . fluid ounces

297.893 679 7 . . . . gills
74.473 419 91 . . . . liquid pints
64 . . . . dry pints*
37.236 709 96 . . . . liquid quarts
35.239 070 166 88 . . . . cubic decimeters or
liters*

32 . . . . dry quarts*
9.309 177 489 . . . . gallons
4 . . . . pecks*
1.244 456 019 . . . . cubic feet
0.221 647 083 1 . . . . petroleum barrel

0.046 090 963 65 . . . . cubic yard
0.035 239 070 166 88 . . . . cubic meter or kiloliter*

One PETROLEUM BARREL is equal to:

2 580 480 . . . . minims*
158 987.294 928 . . . . cubic centimeters or milliliters*

43 008 . . . . fluid drams*
9 702 . . . . cubic inches*
5 376 . . . . fluid ounces*

1 344 . . . . gills*
336 . . . . liquid pints*
288.747 314 5 . . . . dry pints
168 . . . . liquid quarts*
158.987 294 928 . . . . . . cubic decimeters or liters*

144.373 657 2 . . . . dry quarts
42 . . . . gallons*
18.046 707 15 . . . . pecks
5.614 583 333 . . . . cubic feet
4.511 676 789 . . . . bushels

0.207 947 530 9 . . . . cubic yard
0.158 987 294 928 . . . . cubic meter or kiloliter*

One CUBIC YARD is equal to:

12 409 284.16 . . . . minims
764 554.857 984 . . . . cubic centimeters or milliliters*
206 821.402 6 . . . . fluid drams
46 656 . . . . cubic inches*
25 852.675 32 . . . . fluid ounces

6 463.168 831 . . . . gills
1 615.792 208 . . . . liquid pints
1 388.558 514 . . . . dry pints
807.896 103 9 . . . . liquid quarts
764.554 857 984 . . . . . . cubic decimeters or liters*

694.279 257 1 . . . . dry quarts
201.974 026 0 . . . . gallons
86.784 907 13 . . . . pecks
27 . . . . cubic feet*
21.696 226 78 . . . . bushels

4.808 905 380 . . . . petroleum barrels
0.764 554 857 984 . . . . cubic meter or kiloliter*

One CUBIC METER or KILOLITER is equal to:

```
  16 230 730.90  . . . . . . . . . . . . . . . . . . . . . . . . . . minims
   1 000 000  . . . . . . . . . . . . . . . cubic centimeters or milli-
                                                                liters*
     270 512.181 6 . . . . . . . . . . . . . . . . . . . . . . fluid drams
      61 023.744 09 . . . . . . . . . . . . . . . . . . . . cubic inches
      33 814.022 70 . . . . . . . . . . . . . . . . . . . . fluid ounces

       8 453.505 675 . . . . . . . . . . . . . . . . . . . . . . . . . gills
       2 113.376 419 . . . . . . . . . . . . . . . . . . . liquid pints
       1 816.165 969 . . . . . . . . . . . . . . . . . . . . . dry pints
       1 056.688 209 . . . . . . . . . . . . . . . . . . liquid quarts
       1 000  . . . . . . . . . . . . . . . cubic decimeters or liters*

         908.082 984 3  . . . . . . . . . . . . . . . . . dry quarts
         264.172 052 4  . . . . . . . . . . . . . . . . . . . . gallons
         113.510 373 0  . . . . . . . . . . . . . . . . . . . . . pecks
          35.314 666 72  . . . . . . . . . . . . . . . . . cubic feet
          28.377 593 26  . . . . . . . . . . . . . . . . . . bushels

           6.289 810 770  . . . . . . . . . . petroleum barrels
           1.307 950 619  . . . . . . . . . . . . . . . cubic yards
```

ADDITIONAL UNITS OF VOLUME

Metric

1 cubic milli-meter	=	0.000 000 001 . . cubic meter*
	=	0.000 001 . . . . liter*
1 microliter	=	0.000 000 001 . . . cubic meter*
	=	0.000 001 . . . . liter*
1 centiliter	=	0.000 01 . . . . . . cubic meter*
	=	0.01 . . . . liter*
1 deciliter	=	0.000 1 . . . . cubic meter*
	=	0.1 . . . . liter*
1 dekaliter	=	0.01 . . . . cubic meter*
	=	10 . . . . liters*

1 decistere	=	0.1	. . . . cubic meter*
	=	100	. . . . liters*
1 hectoliter	=	0.1	. . . . cubic meter*
	=	100	. . . . liters*
1 stere	=	1	. . . . cubic meter*
			(in tables)
	=	1 000	. . . . liters*
1 dekastere	=	10	. . . . cubic meters*
	=	10 000	. . . . liters*
1 cubic deka- meter	=	1 000	. . . . cubic meters*
	=	1 000 000	. . . . liters*
1 cubic hecto- meter	=	1 000 000	. . . . cubic meters*
	=	1 000 000 000	. . . . liters*
1 cubic kilo- meter	=	1 000 000 000	. . . . cubic meters*
	=	1 000 000 000 000	. . . . liters*

Customary

1 measuring tea- spoon	=	1/3	. . . . . measuring tablespoon*
	=	1 1/3	. . . . drams*
	=	4.928 922	. . cubic centimeters
1 measuring tablespoon	=	0.5	. . . . fluid ounce*
	=	4	. . . . drams*
	=	14.786 76	. . . cubic centimeters
1 measuring cup	=	8	. . . . fluid ounces*
	=	0.236 588 2	. . . . liter
1 board-foot	=	144	. . . . cubic inches*
	=	2.359 737	. . . cubic decimeters
1 dry gallon	=	0.125	. . . . bushel*
	=	4.404 884	. . . cubic decimeters
1 firkin	=	72	. . . . liquid pints*
	=	34.068 71	. . . . cubic decimeters
1 bushel struck measure	=	1	. . . . bushel*
			(in tables)

	=	35.239 070 166 88 . . . cu deci-meters*
1 heaped bushel	=	2 747.715 . . . . cubic inches
	=	45.026 98 . . . . cubic decimeters
1 standard cran-berry barrel	=	86-45/64 . . . . dry quarts*
	=	95.479 30 . . . . cubic decimeters
1 standard barrel for fruits, vegetables, and other dry commodities, except cran-berries	=	7056 . . . . cubic inches*
	=	0.115 627 1 . . . . cubic meter
1 hogshead	=	63 . . . . gallons*
	=	0.238 480 9 . . . . cubic meter
1 cord-foot	=	16 . . . . cubic feet*
	=	0.453 069 5 . . . . cubic meter
1 masonry perch	=	24.75 . . . . cubic feet*
	=	0.700 842 0 . . . . cubic meter
1 tun	=	252 . . . . gallons*
	=	0.953 923 8 . . . . cubic meter
1 measurement ton	=	40 . . . . cubic feet*
	=	1.132 674 . . . . . . cubic meters
1 chaldron	=	36 . . . . bushels*
	=	1.268 607 . . . . . . cubic meters
1 register ton	=	100 . . . . cubic feet*
	=	2.831 684 659 2 . . . . . . cubic meters*
1 firewood cord	=	128 . . . . cubic feet*
	=	3.624 556 . . . . . . cubic meters
1 acre-inch	=	3 630.015 . . . . cubic feet
	=	102.790 6 . . . . cubic meters
1 acre-foot	=	43 560.17 . . . . cubic feet
	=	1 233.487 . . . . cubic meters

REFERENCES

1. BARBROW, LOUIS E., AND JUDSON, LEWIS V. Units and Systems of Weights and Measures; Their Origin, Development, and Present Status. National Bureau of Standards Letter Circular LC 1035. U.S. Department of Commerce. Prepared by: Lewis V. Judson, Office of Weights and Measures, National Bureau of Standards, January 1960. Revised by: Louis E. Barbrow, Office of Metric Information, National Bureau of Standards, June 1975. Identified in the text as: Barbrow and Judson (ref 1).

2. COHEN, E. RICHARD, AND TAYLOR, BARRY N. Fundamental Constants. Compiled by Cohen and Taylor under the auspices of the Task Group on Fundamental Constants of the Committee on Data for Science and Technology (CODATA), International Council of Scientific Unions. Officially adopted by CODATA. Printed in J. Phys. Chem. Ref. Data, volume 2, number 4, page 741 (1973) and CODATA Bulletin number 11 (December 1973). Reprinted in Dimensions, National Bureau of Standards, volume 58, number 1, pages 4 and 5 (January 1974). Identified in the text as: Cohen and Taylor (ref 2).

3. MECHTLY, E. A. The International System of Units; Physical Constants and Conversion Factors. Second Revision. National Aeronautics and Space Administration. NASA SP-7012, 1973. Identified in the text as: Mechtly (ref 3).

4. PAGE, CHESTER H. What Is Weight? American Journal of Physics, volume 43, number 10, pages 920 and 921 (October 1975). Identified in the text as: Page (ref 4).

5. THE INTERNATIONAL SYSTEM OF UNITS (SI). National Bureau of Standards Special Publication 330, 1977 Edition. U.S. Department of Commerce. Identified in the text as: SI (ref 5).

6. ASTM STANDARD METRIC PRACTICE GUIDE. American Society

for Testing and Materials, E380–76, Feb. 1976. Identified in the text as: ASTM (ref 6).

7. SI UNITS AND RECOMMENDATIONS FOR THE USE OF THEIR MULTIPLES AND OF CERTAIN OTHER UNITS. International Organization for Standardization (ISO), ISO 1000, First Edition – 1973. Identified in the text as: ISO (ref 7).

8. HANDBOOK OF CHEMISTRY AND PHYSICS. Published by the Chemical Rubber Company. Fifty-second Edition, 1971–72. Conversion Factors, pages F–242 to F–264. Identified in the text as: Handbook (ref 8).

INDEX

www.ingramcontent.com/pod-product-compliance
Lightning Source LLC
Chambersburg PA
CBHW021931220326
41598CB00061BA/1056